The fastest swimming bird is the gentoo penguin. It swims more than 18½ mph.

Probably the oldest living land animals are the tortoises. The oldest ever recorded was a Marion's tortoise which lived for 152 years.

P9-BJB-565

The smallest seeds are produced by plants called orchids. Orchids living on trees have especially small seeds. They are so small that 1 ounce would contain 40,000,000 seeds.

The Asiatic elephant has the longest gestation time of all mammals. The baby stays inside its mother for an average period of 609 days before it is born. Sometimes this period is even longer and 760 days has been recorded for one baby elephant.

The world's biggest jellyfish is the Arctic giant jellyfish. The diameter of its bell may be 6½ ft and its tentacles stretch 98 ft.

The highest living mammal is the yak. It lives in the Himalayas at a height of 3½ miles or more.

The world's heaviest flying bird is the kori bustard from East and Southern Africa. It sometimes weighs as much as 40 lb.

The biggest spiders' webs are spun by orb weaver spiders. Some webs have been measured with a circumference of 18½ ft.

The longest seaweed is the Pacific giant kelp. It can grow to a length of nearly 66 yd.

The longest insect is the giant stick insect from southeast Asia. It measures 13 in.

The world's biggest forested area is found in Northern Russia. It covers an area of 68,200,000 mi²

The North Atlantic deep sea clam grows more slowly than any other animal. It grows about .32 in in 100 years.

The loudest insect is the male cicada. It can be heard more than 365 yd away.

The world's most poisonous fungus is the death cap toadstool. It can cause death within a few hours of eating it.

↑ The peregrine falcon is the world's fastest animal. In a steep dive it reaches 217 mph.

The biggest land snail is the African giant snail. These snails often have a shell longer than 14 in.

HOW MANY MORE RECORD BREAKERS CAN YOU FIND IN THIS BOOK?

THE LIVING WORLD

THE LIVING WORLD

By Tony Seddon and Jill Bailey

Doubleday & Company, Inc., Garden City, New York

This book was designed and produced by
BLA Publishing Limited,
East Grinstead, Sussex, England.
A member of the Ling Kee Group
LONDON · HONG KONG · TAIPEI · SINGAPORE · NEW YORK

ISBN: 0-385-23754-5

Contents

Note to the reader
On page 153 of this book you will find the glossary. This gives brief explanations of words which may be new to you.

Acknowledgements/Picture Credits

The publishers wish to thank the following people and organizations for their invaluable assistance in the preparation of this book:

Ian Redmond   Paignton Zoo, Devon   Flora and Fauna Preservation Society
ARTISTS:

David Anstey; Fiona Fordyce; Helen Kennett; Karen Moxon; Colin Newman/
Linden Artists; David Parkins; Sallie Alane Reason; Mandy Shepherd;
Rosie Vane-Wright; Phil Weare/Linden Artists; Michael Woods.

PHOTOGRAPHIC CREDITS
t = top; b = bottom; c = centre; l = left; r = right.

COVER: Stephen Dalton/NHPA HALF-TITLE Jen and Des Bartlet/Survival Anglia.
10/11 S.Krasemann/NHPA. 10t Stephen Dalton/NHPA. 10c John and Gillian
Lythgoe/Seaphot. 10b David Rootes/Seaphot. 11t John and Gillian
Lythgoe/Seaphot. 11c Sean Morris/OSF. 11b J.C. Stevenson/OSF. 18 Peter
Johnson/NHPA. 19 Ian Redmond. 23 Mike Price/Survival Anglia. 29 Richard
Matthews/Seaphot. 32t Franz J.Camenzind/Seaphot. 32b Nick Greaves/Seaphot.
33 M.P.L. Fogden/OSF. 34t Barrie E. Watts/OSF. 34b Ken Vaughan/Seaphot. 35 Sean
Avery/Seaphot. 36t Jonathon Scott/Seaphot. 36c Flip Schulke/Seaphot. 37t Peter
David/Seaphot. 37cl Richard Matthews/Seaphot. 37cr Michael Fogden/OSF. 38t Ken
Griffiths/NHPA. 38c Barrie E. Watts/OSF. 39t Alan Root/Survival Anglia. 39c Stephen
Dalton/NHPA. 40t Anthony Bannister/NHPA. 40 c Dr. J.A.L. Cooke/OSF. 40b Anthony
Bannister/NHPA. 41t G.I. Bernard/OSF. 41b Jill Bailey. 42l Bill Wood/NHPA. 42r J.H.
Carmichael/NHPA. 43t Jill Bailey. 43b Bill Wood/NHPA. 44t J.A.L. Cooke/OSF. 44b N.A.
Callow/NHPA. 50l Peri Coelho/Seaphot. 50r Stephen Dalton/NHPA. 51 Karl
Switak/NHPA. 52cl Jill Bailey. 52cr Stephen Dalton/NHPA. 52b John Shaw/NHPA. 53l,
53r Philip Sharpe/OSF. 55 Anthony Bannister/NHPA. 56tl Stephen Dalton/NHPA.
56tr R.J. Erwin/NHPA. 56bl M.W.F. Tweedie/NHPA. 56br Stephen Dalton/NHPA.
57tl Michael Fogden/OSF. 57tr Geoff du Feu/Seaphot. 57c Michael Fogden/OSF. 58 Dr
F. Köster/Survival Anglia. 59 K and K. Ammann/Seaphot. 60 Douglas Dickens.
62 Doug Allan/OSF. 63 Manfred Danegger/NHPA. 65 David C. Fritts/Animals
Animals /OSF. 66 Jonathon Scott/Seaphot. 67tl, 67tr Z. Leszczynski/Animals
Animals/OSF. 67b Peter Parks/OSF. 68 J.A.L. Cooke/OSF.
70 Sean Morris/OSF. 71t Karl Switak/NHPA. 71b David Thompson/OSF.
73 J.B. Free/NHPA. 74 R.J. Erwin. 75 Jonathon Scott/Seaphot. 76 Patrick Fagot/NHPA.
78 Massart/Jacana. 79 Jeff Foot/Survival Anglia. 80 Stephen Dalton/NHPA. 82
Richard Matthews/Seaphot. 85t J.B. Davidson/Survival Anglia. 85b Douglas
Dickens/NHPA. 86 Peter Gathercole/OSF. 87 J.A.L. Cooke/OSF. 88 Stephen
Dalton/NHPA. 92 Jill Bailey. 93t Leonard Lee Rue III/Animals Animals/OSF. 93b Jeff
Foot/Survival Anglia. 95tr G.J. Cambridge/NHPA. 95c John Shaw/NHPA. 95b Stephen
Dalton/NHPA. 96t J.A.L. Cooke/OSF. 96b Jill Bailey. 97t Avril Rawage/OSF. 97b Ken
Lucas/Seaphot. 98 John Shaw/NHPA. 99r S. Krasemann/NHPA. 99l Peter
Stephenson/Seaphot. 100t S. Krasemann/NHPA. 100b Brian Hawkes/NHPA. 101 Jack
Lentfer/Survival Anglia. 102tl John Shaw/NHPA. 102tr Brian Milne/Animals
Animals/OSF. 102c S. Krasemann/NHPA. 103 Hellio and Van Ingen/NHPA. 104t Jeff
Foot/Survival Anglia. 104b Jen and Des Bartlett/Survival Anglia. 105 David and Sue
Cayless/OSF. 106t E. Hanumantha Rao/NHPA. 106b Martyn F. Chillmaid/OSF.
107 Stephen Dalton/NHPA. 108t Michael Leach/NHPA. 108b Marty Stouffer
Productions Ltd/Animals Animals/OSF. 109t Jonathon Scott/Seaphot. 109b John
Paling/OSF. 110 D.H. Thompson/OSF. 111t Jack Wilburn/Animals Animals/OSF.
111b S. Robinson/NHPA. 114t Jim Standen. 114b, 115 Ivan Polunin/NHPA. 117t Peter
Parks/OSF. 117b David Maitland/Seaphot. 118 Peter Parks/OSF. 119t Peter
David/Seaphot. 119b Robert Hessler/Seaphot. 120t Warren Williams/Seaphot.
120b Hugh Jones/Seaphot. 121 Linda Pitkin/Seaphot. 123 Melvin Grey/NHPA.
124 Stephen Dalton/NHPA. 124/125 A. C. Waltham. 126l Peter Johnson/NHPA.
126r Jeff Foot/Survival Anglia. 127l Ralph and Daphne Keller/NHPA. 127r Raymond
Blythe/OSF. 128 J.B. Blossom/NHPA. 129 L. Campbell/NHPA. 136 Tony Morrison/South
American Pictures. 137 E. Hanumantha Rao/NHPA. 138t Patrick Fagot/NHPA.
138b Jeremy Cherfas. 139 Paignton Zoo Educational Service. 140, 141 Jill Bailey.
142t C.C. Lockwood/Animals Animals/OSF. 142c David W. Macdonald/OSF.
142b, 143tl G.I. Bernard/OSF. 143tr James Carmichael/NHPA. 143b J.A.L. Cooke/OSF.

Introduction

Did you know that humans share the Earth with more than 3,000,000,000,000,000,000,000,000,000,000,000 other living things. You probably have difficulty imagining how big this number is. It is a very large number indeed. If you counted once every second it would take you more than one hundred million, million, million, million years to reach the total.

This book tells you something about some of the one million or so different species of animals and plants which make up this enormous number of living things. It gives you lots of information, but it also encourages you to ask questions about the living world. Perhaps you are already well known for asking questions. It is good to ask "Why?" especially when you look at animals and plants. So keep on asking and try to find the answers.

A famous scientist once said:
> The *is-ness* of things is well worth studying but
> it's their *why-ness* that makes life worth living.

By the time you get to the end of this book a lot of your "Whys?" will have been answered, but you will still have more. In fact, we hope this book makes you ask more questions than it answers. This is the way you will learn more about the one million or so different species of animals and plants who are your neighbors on the Earth.

AROUND THE WORLD NOTEBOOK

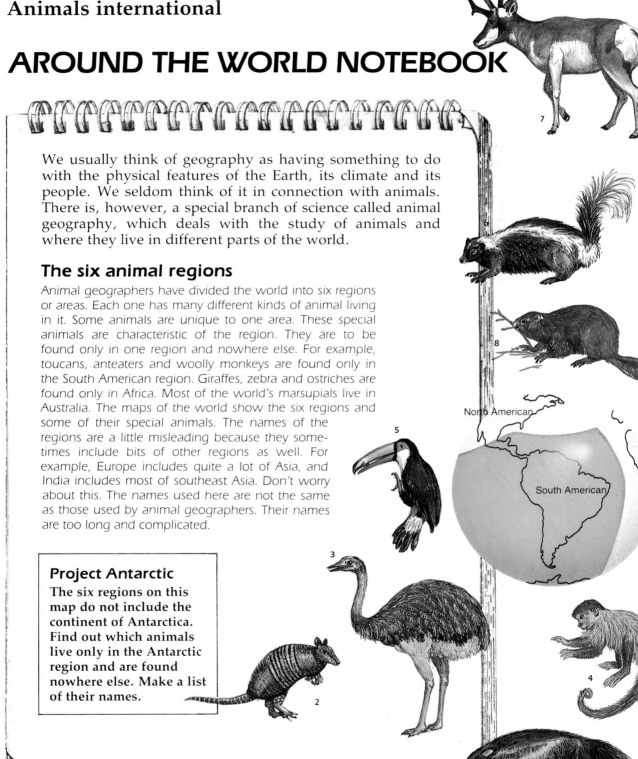

We usually think of geography as having something to do with the physical features of the Earth, its climate and its people. We seldom think of it in connection with animals. There is, however, a special branch of science called animal geography, which deals with the study of animals and where they live in different parts of the world.

The six animal regions

Animal geographers have divided the world into six regions or areas. Each one has many different kinds of animal living in it. Some animals are unique to one area. These special animals are characteristic of the region. They are to be found only in one region and nowhere else. For example, toucans, anteaters and woolly monkeys are found only in the South American region. Giraffes, zebra and ostriches are found only in Africa. Most of the world's marsupials live in Australia. The maps of the world show the six regions and some of their special animals. The names of the regions are a little misleading because they sometimes include bits of other regions as well. For example, Europe includes quite a lot of Asia, and India includes most of southeast Asia. Don't worry about this. The names used here are not the same as those used by animal geographers. Their names are too long and complicated.

Project Antarctic

The six regions on this map do not include the continent of Antarctica. Find out which animals live only in the Antarctic region and are found nowhere else. Make a list of their names.

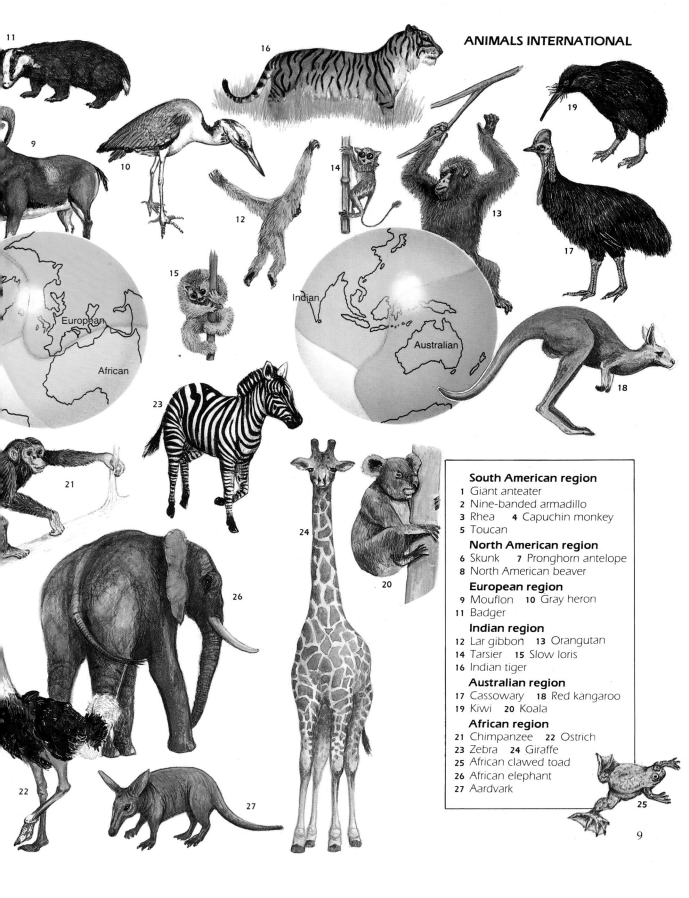

South American region
1 Giant anteater
2 Nine-banded armadillo
3 Rhea   4 Capuchin monkey
5 Toucan

North American region
6 Skunk   7 Pronghorn antelope
8 North American beaver

European region
9 Mouflon   10 Gray heron
11 Badger

Indian region
12 Lar gibbon   13 Orangutan
14 Tarsier   15 Slow loris
16 Indian tiger

Australian region
17 Cassowary   18 Red kangaroo
19 Kiwi   20 Koala

African region
21 Chimpanzee   22 Ostrich
23 Zebra   24 Giraffe
25 African clawed toad
26 African elephant
27 Aardvark

Plants international

Plants around the world

Plants are found all over the world, except near the North and South Poles, on the tops of high mountains, and in very dry deserts. Every plant species has its own special needs for temperature, light, water and mineral salts. Plants grow where the conditions suit these needs. So different plants grow in different parts of the world. If there were no farms and cities, or other kinds of human interference, certain types of wild or 'natural' vegetation would occur in particular parts of the world. Here you see the main types of natural vegetation in the northern hemisphere. Similar types of plants are found in the southern hemisphere in places where the climate and soil are suitable.

DECIDUOUS FOREST
Broad-leaved trees which lose their leaves in winter. Flowering herbs on forest floor. Shrubs, brambles and other scrambling and climbing plants.

WEST

MEDITERRANEAN FOREST
Broad-leaved trees, many of them evergreen, with waxy, shiny leaves. Many flowering herbs and shrubs, often thorny.

RAIN FOREST
Dense forest with many evergreen trees. Farther from the Equator there are more trees that shed their leaves from time to time. Many vines and creepers. Epiphytic plants (plants that grow on branches and trunks of trees). Many forest herbs and ferns.

Plants from north to south

In the far north, low temperatures prevent plants growing very tall. Farther south it is warmer, and there are forests. First, there are evergreen conifer forests whose trees can cope with poor soils and cold, snowy winters. As the climate gets warmer and wetter, we find forests of trees which shed their leaves in winter. Still farther south, the summers are hot and dry. Here the forests are mainly of trees with shiny, waxy leaves that do not lose water easily. Finally there are the wet, warm tropical regions with their lush rain forests.

H

TUNDRA
Mosses, lichens, sedges. Slow-growing cushion plants, very short. Small evergreen dwarf shrubs.

Plenty of variety

Throughout the world there are patches of other types of vegetation, such as bogs, swamps and salt marshes, where local conditions are rather special. High mountains also create a range of different conditions for plants. Even in the tropics, the vegetation will change with altitude (height), from tropical forest to coniferous forest, and, eventually, tundra-like vegetation on the cold, exposed mountain tops.

CONIFEROUS FOREST
Mostly evergreen trees with needlelike leaves. A few broad-leaved trees like birch, willow and alder. Few flowering herbs on forest floor because of the deep shade there.

Plants from west to east

In the northern latitudes, the winds blow mainly from the west. They bring moist air from the oceans to the land. As you go farther inland, there is less rainfall. The land heats up and cools down faster than the sea. So areas inland have hotter summers and colder winters than those near the coast. The forest gives way to grasslands and scrub (small bushes) and then desert.

EAST

GRASSLAND
Includes the European steppes and the American prairies. Mainly grasses, often forming tufts or clumps. Some shrubs and stunted trees. Many flowering herbs, mostly springing up from seeds or underground storage organs after rain.

DESERT
Lots of bare ground with scattered herbs, shrubs and small trees. Fleshy plants — cacti and plants with swollen leaves and stems for storing water, and spines for protection. Many plants which grow from seed after rain, flower and produce seeds in a few weeks, then die.

Animals with backbones

You have to be a detective to classify an animal. You look for clues.

Animal groups

Scientists are quite tidy people. If they can, they like to put things into groups. They do this with animals. They look for patterns which we call characteristics.

Animals with similar characteristics are put in the same group. This is called classification. The backboned animals are called vertebrates, and there are about 45,000 different species alive today. Scientists classify vertebrates into five main groups or classes. These are fish, amphibians, reptiles, birds and mammals. Each class can be divided further into smaller groups called orders. For example, mammals which eat meat are grouped in the carnivore order. Even orders can be divided into smaller groups. You can learn about the smallest group on pages 146 and 147.

Classifying an animal can be difficult. Animals are not always what they seem to be at first sight. For example, whales and dolphins look like fish, but a closer look shows them to be mammals. The pangolin lives in Africa and Asia. Its body is covered with scales. Do you know which vertebrate class the pangolin belongs to? Answer on page 152.

What is a fish?

There are about 23,000 different species of fish alive today.

Fish:
○ are cold-blooded
○ live in water
○ breathe by gills
○ are covered in scales.

There are three main groups: the jawless fish (hagfish and lampreys), the cartilaginous fish whose skeleton is made of cartilage (sharks and rays), and the bony fish whose skeleton is made of bone.

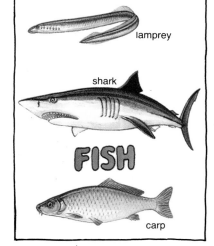

lamprey

shark

FISH

carp

What is an amphibian?

There are more than 4,000 different species of amphibians alive today.

Amphibians:
○ are cold-blooded
○ have smooth, wet skins
○ return to water to breed
○ breathe with gills when young, but as adults use their skins and lungs.

There are three main groups. These are the legless caecilians, the newts and salamanders, and the frogs and toads.

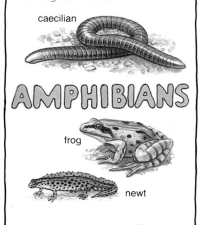

caecilian

AMPHIBIANS

frog

newt

What is a reptile?

There are about 5,200 different species of reptile alive today. The majority of these are snakes and lizards.

Reptiles:
○ are cold-blooded
○ breathe air using lungs
○ have dry, scaly skins

There are four main groups. These are the turtles and tortoises, the lizards, the snakes and the crocodiles and their relatives.

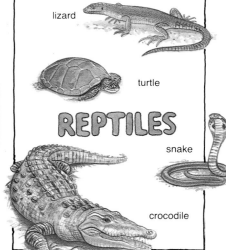

lizard

turtle

REPTILES

snake

crocodile

What is a bird?

There are about 8,600 different species of bird alive today.

Birds:
o have wings
o are warm-blooded
o breathe air using lungs
o have feathers
o have no teeth
o lay eggs.

Not all birds fly. Some have wings which are too small for flight. Penguins have flippers instead of wings. Ostriches have long legs for fast running. There are 27 orders of birds divided into different families. Below are some examples of different birds.

kiwi

penguin

ostrich

owl

goose

BIRDS

What is a mammal?

There are about 4,000 different species of mammal alive today.

Mammals:
o are warm-blooded
o breathe air using lungs
o have body hair
o produce milk and suckle their young.

Altogether there are 19 orders of mammals. Below are some of these.

gibbon

Lemurs, monkeys and apes, or primates, e.g. spider monkeys, ring-tailed lemurs and chimpanzees.

dolphin

Whales and dolphins. Mostly live in the sea but some dolphins live in muddy rivers.

MAMMALS

Egg-laying mammals, e.g. spiny anteaters and duck-billed platypuses.

spiny anteater

Gnawing mammals or rodents, e.g. rats and mice. Nearly half the living species of mammals are rodents.

mouse

Pouched mammals or marsupials, e.g. kangaroos, wombats.

kangaroo

Meat-eaters or carnivores, e.g. lions, wolves and hyenas.

lion

Odd-toed hoofed animals, e.g. horses, tapirs and rhinoceroses.

tapir

Insect-eaters or insecti-vores, e.g. shrews.

shrew

Even-toed hoofed animals, e.g. giraffes, camels and hippopotamuses.

hippopotamus

Flying mammals or bats. These are the only true flying mammals.

bat

Animals without backbones

ECHINODERMATA
○ Animals with tough
  spiny skins.
○ Their bodies are
  designed so that parts
  are arranged in fives or
  multiples of five (except
  sea cucumbers).

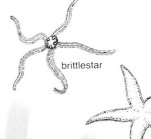

brittlestar

sea urchin

VERTEBRATES

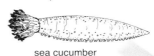

starfish

sea cucumber

ECHINODERMATA

CHORDATA

sea lily

The invertebrates

Scientists classify animals,
they put animals with
similar characteristics into
groups. Animals without
backbones are a very large
and varied group. They are
called the invertebrates.
Ninety-five percent of all
animals are invertebrates.
There are about 950,000
species of invertebrates.
They range from micro-
scopic animals, too small
for the eye to see, to the
giant squid, which may be
up to 65 feet long. They
can be found in water, on
land, in the soil, in the
air, and even in the snow
of the polar ice caps.

Here you can see a few
of the more common
groups of invertebrates
and their international
scientific names. Beside
each group are some clues
for recognizing them.

jellyfish

COELENTERATA
○ Soft, jellylike animals
  with hollow bodies.
○ They catch prey on
  stinging tentacles.

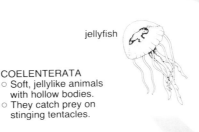

COELENTERATA

sea anemones and corals

hydra

PORIFERA
○ Animals with stiff bodies.
○ Sponges filter-feed by
  passing water through
  their bodies.
○ Water enters and leaves
  through a series of holes
  called pores.

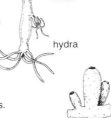

sponge

PORIFERA

PROTOZOA
○ Very simple animals
  made up of just one cell.
  Most are no bigger than
  0.04 in wide.

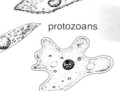

protozoans

PROTOZOA

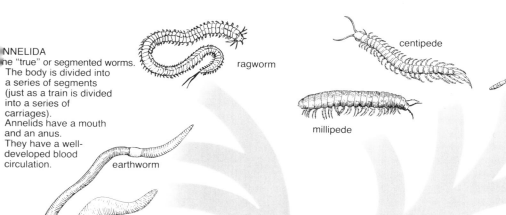

centipede

insect

millipede

ANNELIDA
The "true" or segmented worms. The body is divided into a series of segments (just as a train is divided into a series of carriages). Annelids have a mouth and an anus. They have a well-developed blood circulation.

ragworm

earthworm

leech

spider

ANNELIDA

ARTHROPODA

crab

BRACHIOPODA
Lamp shells have a pair of shells joined by a hinge. They filter-feed using a ring of tentacles in between the two shells.

lamp shell

BRACHIOPODA

MOLLUSCA

ARTHROPODA
○ Animals with jointed legs.
○ The body is covered in a hard skin or shell called the cuticle.
○ The body is divided into segments.

MOLLUSCA
○ Soft-bodied animals, often surrounded by a hard shell or shells. Squids and octopuses may have the shell inside the body.
○ The internal organs are usually covered with a mantle — a folded sheet of tissue.

land snail

octopus

sea mat or moss animal

ECTOPROCTA

ECTOPROCTA
○ Tiny animals that live in colonies enclosed in stony tubes or sheets.
○ Moss animals cannot move, but they filter-feed using a circle of tentacles.

NEMERTINA
○ Flat-bodied, very long worms, up to 90 ft long.
○ Ribbonworms are similar to flatworms, but with an opening at both ends of the gut — the mouth and the anus.
○ They have a simple blood circulation.

mussel

ribbon worm

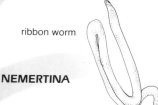

NEMERTINA

flatworms

PLATYHELMINTHES

PLATYHELMINTHES
○ Simple animals whose gut has only one opening: the mouth.
○ Flatworms have no blood circulation.

A third kingdom?

Until recently, scientists have divided the living world into two kingdoms, the plant kingdom and the animal kingdom. Today, many scientists put the protozoa and some of their relatives in a third kingdom, called the Protista or Protoctista. The cells of protozoa are very different from those of other animals. They are much more complex, since they carry out all the activities of the animal.

Plants and other nonanimals

What is a plant?

People used to think that living organisms could be divided into just two main groups — the Plant Kingdom and the Animal Kingdom. It is easy to suppose that if a living creature is not an animal, it must be a plant. But that depends on what you think a plant is.

Plants are independent

The main difference between plants and animals is that plants can use simple chemicals found in the air, water and soil to make all the materials needed to build their bodies, while animals have to eat plants or other animals.

Plants have special cells

Another feature of plants is that their cells — the smallest units of living material — are surrounded by walls made of a substance called cellulose.

Definitely plants

There are several groups of nonanimals which are definitely green plants. They include the mosses and liverworts, the ferns, clubmosses and horsetails, the conifers, ginkgoes and cycads, and the flowering plants. All of them use the green substance chlorophyll to make their own food by photosynthesis.

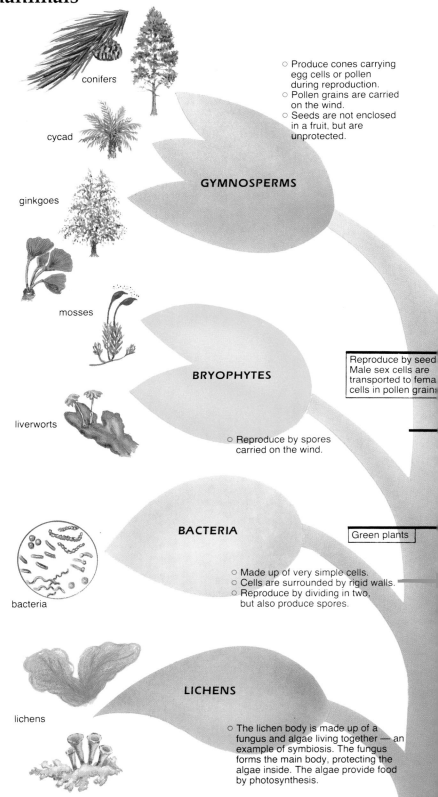

conifers

cycad

ginkgoes

mosses

liverworts

bacteria

lichens

GYMNOSPERMS

○ Produce cones carrying egg cells or pollen during reproduction.
○ Pollen grains are carried on the wind.
○ Seeds are not enclosed in a fruit, but are unprotected.

BRYOPHYTES

Reproduce by seed Male sex cells are transported to fema cells in pollen grain:

○ Reproduce by spores carried on the wind.

BACTERIA

Green plants

○ Made up of very simple cells.
○ Cells are surrounded by rigid walls.
○ Reproduce by dividing in two, but also produce spores.

LICHENS

○ The lichen body is made up of a fungus and algae living together — an example of symbiosis. The fungus forms the main body, protecting the algae inside. The algae provide food by photosynthesis.

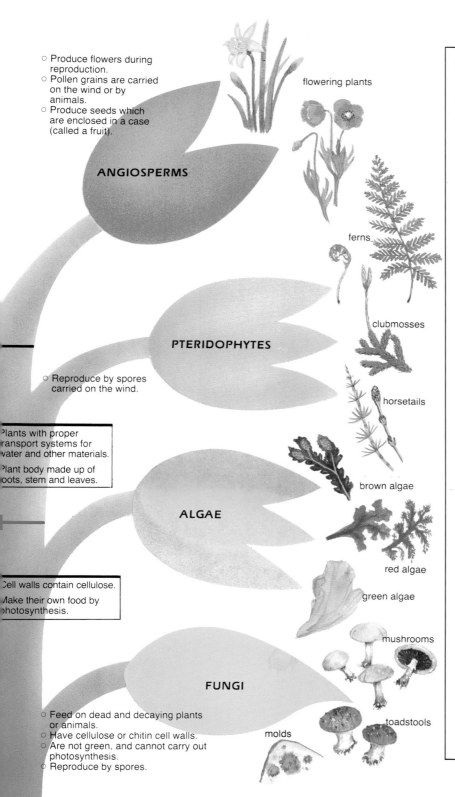

○ Produce flowers during reproduction.
○ Pollen grains are carried on the wind or by animals.
○ Produce seeds which are enclosed in a case (called a fruit).

ANGIOSPERMS

flowering plants

ferns

PTERIDOPHYTES

clubmosses

○ Reproduce by spores carried on the wind.

horsetails

Plants with proper transport systems for water and other materials.
Plant body made up of roots, stem and leaves.

brown algae

ALGAE

red algae

green algae

Cell walls contain cellulose.
Make their own food by photosynthesis.

mushrooms

FUNGI

toadstools

molds

○ Feed on dead and decaying plants or animals.
○ Have cellulose or chitin cell walls.
○ Are not green, and cannot carry out photosynthesis.
○ Reproduce by spores.

What do you call a nonanimal?

There are some living creatures that do not appear to fit in either the Plant Kingdom or the Animal Kingdom. For instance, what do you call a living organism that is rooted to the ground, has a branching body and has cellulose cell walls, but is not green, and cannot make its own food?

This is a description of a fungus. Most scientists classify the fungi as a separate group of living creatures — the Kingdom Fungi. In some fungi the cell walls are made of chitin, a substance found in insects.

Bacteria have much simpler cells than other living creatures, and may also be placed in a kingdom of their own, the Kingdom Monera.

Because plants and other nonanimals have soft bodies and do not leave many fossils, we do not know how they are related to each other. Scientists have different ideas about classifying them. Here we have simply shown you the main groups, without trying to put them in a special order or giving them their special international names.

Variety is the spice of life

What is a species?

There are more than one million different types of animals and plants living on the Earth. Each type is called a species. A species is a group of animals or plants which look similar and which breed together to produce fertile offspring. So, for example, a lion is a different species from a tiger. They look very different and are easy to tell apart. But could you tell one lion from another, or could you distinguish one tiger from a group of several?

Every animal and plant is unique

Unless you are one of a pair of identical twins, there is nobody else in the world who looks exactly like you. You are a unique animal and so is every other animal. Everybody you know looks different, and you have no difficulty telling one person from another. This is true, even though all humans belong to the same species. Just as members of the human species vary, so do members of other species. No two giraffes are identical, although at first sight you might think so. Scientists call these differences variations. These variations can be big or small. Sometimes you have to look very carefully to see them.

Each member of a population of oak trees in a wood is different from its neighbors. Trees of the same age are different in height and girth. No two trees have the same number of leaves and there is a variety of leaf shapes and sizes. The number of acorns each tree produces varies. Even acorns from the same tree vary in size and weight. All these variations are difficult to see, but they are there. You can easily distinguish between your friends. Can you tell one oak tree from another?

There are many varieties of dogs but they all belong to the same species. They vary in size and shape because they have been bred for different jobs. Humans have chosen different characteristics over hundreds of years to produce the variety of dogs we know today. Even the odd-shaped dachshund was bred for a reason. Can you guess what it was? And why do you think the bulldog looks like it does? Answers on page 152.

▲ At first sight the members of this herd of zebra look identical. But are they? Look closely at the pattern of stripes on each animal. Is this pattern exactly the same on each zebra? Choose four zebra and try to draw the pattern made by their stripes. Compare your drawings. Are they all the same? Some scientists studying zebra populations become so familiar with the group that they are able to tell one zebra from another. They identify the pattern of stripes.

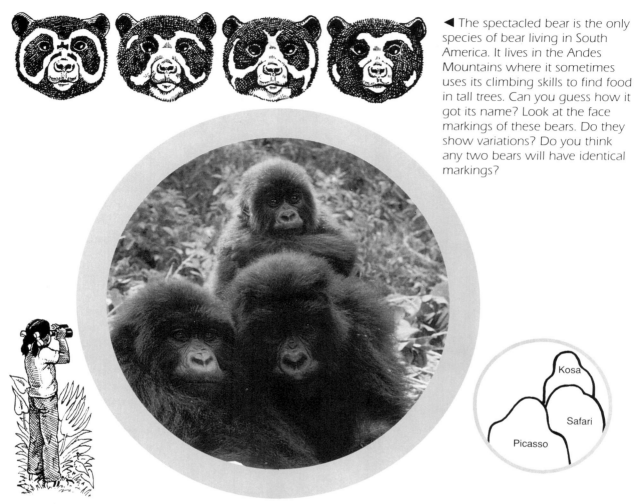

◄ The spectacled bear is the only species of bear living in South America. It lives in the Andes Mountains where it sometimes uses its climbing skills to find food in tall trees. Can you guess how it got its name? Look at the face markings of these bears. Do they show variations? Do you think any two bears will have identical markings?

Kosa

Safari

Picasso

Noseprint identikit

Dian Fossey, an American naturalist, studied the wild mountain gorilla from tropical Africa. She got to know the population so well she was able to recognize individuals. She gave each gorilla a name. Each mountain gorilla has its own individual-shaped nose. No two gorilla noses are the same. Just as humans have unique fingerprints, gorillas have "noseprints." Dian Fossey was able to identify each gorilla in the population by its "noseprint." Can you do the same? Above is a photograph of a group of gorillas. You can find out their names by looking at the small outline drawing. Left is a drawing of one of these gorillas. Look at its "nose-print." Is it Picasso, Kosa or Safari? Answer on page 152.

▲ Is this the face of Picasso, Kosa, or Safari?

Evolution and change

More than enough offspring

You have already read something about variation on pages 18 and 19. There is more on this topic later on this page. But first some mathematics. A famous naturalist called Charles Darwin once worked out that, after seven hundred years, a pair of African elephants could have nineteen million descendents. The cockroach is even more remarkable. After only seven months, a pair of cockroaches could have 164 billion descendents. How can this be? Why isn't the world covered in elephants and cockroaches? To find the answer to the first question, look at the chessboard game. To answer the second, look at the pictures below. Animals are capable of increasing in numbers like the coin increases in value in the chessboard game. However, they don't. The pictures help you to understand why.

Try this on your friends

Put a nickel on the chessboard as shown. Move the coin from one square to the next. Imagine that the coin doubles in value every time it lands on a new square. So, for example, it is worth 5¢ on square 1, 10¢ on square 2, 20¢ on square 3, and so on. Ask your friends to guess how much money you have "made" when you get to the last square at the top of the board. You might surprise them.

1
One pair of rabbits can produce a litter of babies six times a year.

2
If all the babies survived and continued to breed the world would be overrun with rabbits.

3
Scientists have found out that animal populations don't change much in size. This is because not all the offspring survive or breed.

4
Rabbits may die of disease
or be eaten by foxes
or not get enough food
or not find a mate
or not find a nesting site.

5
The rabbits in a population are not identical. They differ from each other: they vary. Some of these differences are important for survival.

6
Some rabbits might run faster to escape predators. Others might have coat colors which give camouflage from predators. They are better adapted for survival.

7
The rabbits which survive are better suited to their environment. Those which survive breed to produce offspring like themselves. The babies may inherit the characteristics of speed and coat color. This gives them a better chance of survival. Can you now see why variation in a species is important? This survival of animals suited to the environment is called natural selection. Can you explain why the world isn't overrun by elephants or cockroaches?

What makes variation happen?

How does variation come about? How do different species develop? Both these questions worried Charles Darwin. They have worried many other scientists since Darwin. We know that animals and plants inherit characteristics from their parents. These are passed on in the sperms from the father, and eggs from the mother. Organisms don't all receive exactly the same information from their parents. It varies from one organism to another. It is this different information which offspring inherit which brings about variation. Think again about the spectacled bears on page 19. Can you now explain why they have different face markings? Answer on page 152.

But how do different species come about?

We can't go back in history, so we can never be sure how different species formed. The process usually takes place over very long periods of time. Perhaps many hundreds of thousands of years. We know it involves natural selection. However, scientists also think other things are important as well. There are three stages in species formation.

1

A population of animals or plants may become divided in two by a barrier such as a range of mountains. This stops the members of the two groups from breeding together.

2

Each of the two new groups now begins to adapt to its own local environment on either side of the mountain range. Again natural selection is at work.

The two populations gradually become different from each other. This takes a long time.

3

The two populations may become so different that they may no longer be able to breed together, even if they could meet up again. They have become different species.

Remember

Conditions on the two sides of the barrier will probably be different. The two populations have to adapt to different environments. Natural selection gradually changes the two populations to make them different. We call this evolution.

The Galápagos project

There are 13 species of finch on the Galápagos Islands in the Pacific Ocean. Each has a different shape or size of beak. Below are three examples.

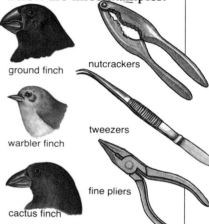

ground finch

nutcrackers

warbler finch

tweezers

cactus finch

fine pliers

Each beak has a common tool drawn alongside to help you understand how it works. When the islands were first formed there were no animals and plants living there. Look at the map of the world and decide where the first finches came from. Find the Galápagos Islands by tracing your finger along the Equator until you come to them. Can you explain why the finches have different beaks? What is the barrier in this case? Look again at the beaks and the three foods at the top of the page. Can you match each beak with the correct food? Answers on page 152.

More about variation (pp.18, 19); reproduction (pp.66, 67)

Animals in armor

Protection and defense

Animals have many ways of protecting and defending themselves. They use color, changes in shape and size, stings and bites, and they sometimes even pretend to be dead. Mollusks such as snails carry a hard shell around with them which protects the soft, delicate body structures inside. Other animals have evolved a rather special way to protect their bodies. They are covered in a suit of armor.

Animal armor

Imagine walking around all day in a suit of armor. You would probably find it very heavy, uncomfortable and difficult to move in. Animal armor is not heavy because it is made of light material. It also has lots of joints and hinges which make it flexible and easy to walk in.

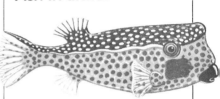

▲ Millipedes are cylindrical animals covered by a tube of armor. The joints between each piece help to make the body very flexible. A millipede can even curl up when attacked.

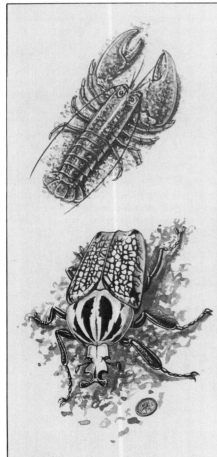

Insects and their relatives — spiders, crabs and lobsters — wear a complete covering of armor. The armor is made of a lightweight substance, called chitin, which is very tough and waterproof. Most insects fly, so they must not be too heavy. The armor not only protects the insect's body, it also supports it like a skeleton. Because it is waterproof, it stops the insect from drying out. At certain points in the armor there are pads of a rubberlike material called resilin. These work like miniature shock absorbers. Every so often an insect throws off its suit of armor so that it can grow. It soon covers itself with a new suit.

Fish in armor

Armor plating is not found in many fish. This is because a fish needs a flexible body for swimming. However, the trunk fish has bony pieces in its skin which fit together to form a box. It can only swim slowly and it cannot wiggle its body like other fish. It has to paddle along with its fins.

The sea horse is also covered in armor. It moves very slowly, swimming upright by means of its gently waving fins.

▲ Reptiles are covered in a scaly skin. In crocodiles and their relatives, the alligators, these scales become thickened, especially on the back and along the tail. They form patches of armor. Crocodiles use their tail for swimming, so its covering of armor plates is jointed to make it flexible. The crocodile moves its tail to push itself through the water.

▲ Tortoises and turtles have a dome-shaped shell on their backs. It is joined to the backbone and forms a rigid covering. The legs are less heavily armored to help in movement. Everything can be pulled under the shell for extra protection.

Pangolins live in Africa and Asia. They are nocturnal animals and are the only mammal whose body is covered in scales. The scales are made from hair. Pangolins walk on their back legs and balance on their tails. They feed on ants and termites, which they dig up with their powerful front claws. Their armor plating gives protection from bites and stings. Pangolins curl into a ball when attacked.

Knights in armor

There are seven species of armadillo living in Central and South America. The nine-banded armadillo has even invaded North America, where people think of it as a pest. The armadillos are the real armor wearers of the animal kingdom. Their protective suits are jointed just like the armor of a medieval knight. The three-banded armadillo rolls into a ball when frightened. Its head and tail fit snugly to complete the protection.

Did you know?

The female nine-banded armadillo always gives birth to identical quadruplets.

A flea has rubber pads of resilin on its back feet. They act like a catapult to shoot the animal into the air over a distance as much as 200 times its own length.

A female pangolin protects its baby by curling up around it.

More about animal defenses (pp.54, 55)

Shaping up to things

Variety of shape

Animals and plants come in many different shapes and sizes. Think of the variety of shapes of living things. Some animals are shaped like a cylinder, or a slightly altered version of a cylinder. Others are very flat, and look as though they have been squashed by a heavy weight. There are animals with a thin body shape. They look as though they have been flattened from either side. Some animals are disk-shaped, some animals are spherical. Many mollusks have a spiral-shaped shell. Other animals have an even more peculiar shape. For example, starfish which, as their name suggests, are shaped like a star. Many animals have different shapes at different times in their life cycles. Others can change their shape at particular moments. Plants also come in different shapes. Some are cylindrical, some spherical, some cone-shaped, and some are flat. Often the different shapes of animals and plants reflect how they have adapted to their different habitats.

Adding to shape

In addition to these basic body shapes, there are structures which give variety to animal and plant shapes. Animals have wings, legs, flippers, fins, tails and flukes (a flattened tail fin). Some animals have horns or antlers. Others have flaps of loose skin such as wattles (fleshy outgrowths) and dewlaps (loose, hanging skin, like that under the throat of a turkey). Plants have a variety of leaf shapes. Some have no leaves, others have spines and thorns.

▲ Cylinder-shaped bodies are common in animals. Often animals which burrow have this shape. So do some animals which live in trees. Snakes are really just long cylinders. So are earthworms and millipedes. The cylindrical shape helps them to move in narrow tunnels underground, and also helps them wiggle between plant stems on the ground.

Thin animals

Many animals have a shape as if they have been squeezed from either side. This is a useful shape for slipping between small gaps. Fish living among coral reefs are this shape, and so are many tree-dwelling chameleons. The chameleons move easily between leaves and twigs. Fleas are also flattened from side to side so they can move between the hairs and feathers of their hosts. Some animals are shaped as though they have been flattened by a heavy weight. Animals living on rocky, exposed surfaces and on the seabed are like this. It makes them less likely to be knocked off by the moving water. Parasitic ticks are flat so that they stay on their host.

▼ This cheetah is running at full speed. When running fast, its shape becomes streamlined. By doing this its body offers less resistance to air as it moves forward. Fast-swimming aquatic animals are also streamlined or torpedo-shaped for easy movement through water. Birds take on a streamlined shape when flying. This helps reduce drag, and allows the bird to move more quickly. Many animals change the position of their limbs to become more streamlined for fast forward movement.

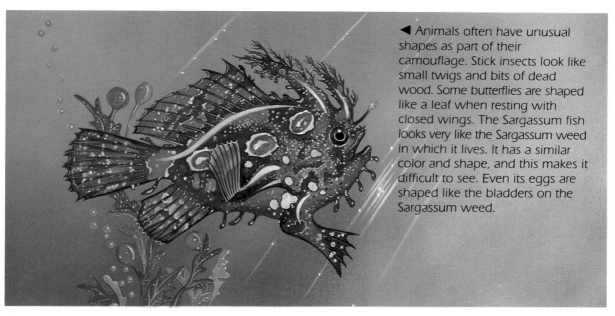

◀ Animals often have unusual shapes as part of their camouflage. Stick insects look like small twigs and bits of dead wood. Some butterflies are shaped like a leaf when resting with closed wings. The Sargassum fish looks very like the Sargassum weed in which it lives. It has a similar color and shape, and this makes it difficult to see. Even its eggs are shaped like the bladders on the Sargassum weed.

Shape for defense

Some animals change shape when attacked. A frog swells up to look bigger. A cat makes its fur stand upright when threatened by a dog. This makes it look bigger in an attempt to frighten the dog away. Some animals curl up into a ball when defending themselves. Hedgehogs do this, and so do armadillos. Many millipedes change from a cylindrical shape to a coiled-disk shape when touched. If attacked, the frilled lizard from Australia will sometimes change its shape by unfolding a huge frill around its neck. This makes its head look bigger and fiercer.

▲ Many animals change shape during courtship, or when defending territory. The male frigate bird blows up a huge throat sac to attract its mate. This makes the front of the bird look like a large, red balloon. The male anole lizard uses a similar device. Its inflated orange throat sac makes it look bigger and more frightening to other males when defending territory, and also attracts a female.

---Did you know?---

Of all the shapes we know, a sphere has the smallest amount of surface compared with its volume. Some cacti living in hot, dry conditions are spherical. Plants lose water from their surface. Because these cacti are spherical, the amount of surface through which water can pass is reduced. This helps the cacti to keep water inside their tissues.

In cold conditions many animals become spherical. Sleeping dogs or cats curl up into a ball. Hibernating mammals curl up to sleep. Can you think why? Answer on page 152.

More about changing shape (pp.68, 69); camouflage (pp.52, 53); courtship (pp.64, 65)

The importance of size

What is big?

What do we mean by big? The biggest living thing is a tree, the giant sequoia. It is many million times bigger than the smallest living organism. Humans are big compared to most living things. Ninety-nine percent of all living things are smaller than humans.

Small and big

If you play with a mouse in your hands, it may escape and fall to the ground. It won't hurt itself if it does this. But a human who fell a distance many times his or her own height would probably die. Small and large bodies of a similar shape and substance have different properties. Small animals have a very big surface compared to their volume. Big animals have a relatively small surface compared to their volume. Look at the square animals in the picture to see if this is true.

Gravity

The reason we don't float off the Earth into space is because of gravity. Gravity is a force which pulls your body back to Earth. All objects are attracted to the Earth by gravity. The body of a big animal contains more material than the body of a small animal. We say it has a bigger mass. The bigger the animal, the stronger the pull of gravity on its body. This is because there is more mass to pull on. Small animals are pulled less by gravity because they have less mass to pull on. Small animals have a large surface compared to their volume. This means they are affected more by surface forces than by gravity. Forces such as wind and rain affect small animals much more than big animals. Does this help to explain the case of the falling mouse and human? Answer on page 152.

Floating in air

Tiny animals and plants have a huge surface compared to their volume. They also have very little mass. Because of this, movements of the air can keep them afloat. Tiny insects, spiders, fungal spores and pollen grains float in air because they are small.

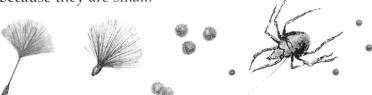

This is a cube-shaped animal. Each side of its body is 1 in long. Its total surface is $6\,in^2$. Its volume is $1\,in^3$. It has $6\,in^2$ of surface for every $1\,in^3$ of volume.

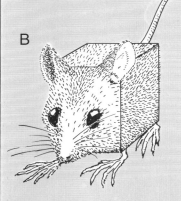

This is a bigger cube-shaped animal. Each side of its body is 6 in long. Its total surface is $216\,in^2$. Its volume is $216\,in^3$. It has $1\,in^2$ of surface for every $1\,in^3$ of volume.

Animal A is smaller than animal B. But it has a much bigger surface compared to its volume than animal B.

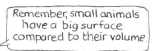

Remember, small animals have a big surface compared to their volume.

What limits size?

The size of an animal depends upon a number of things. One very important factor is how the animal breathes. Insects are quite small animals, and they get their oxygen through tiny holes in the surface of their bodies. The goliath beetle is the biggest insect, but it only weighs about 3.5 ounces. This is about as big as an insect can grow. Any bigger and it would be unable to get all the oxygen it needed. Bigger animals need special breathing equipment like gills and lungs.

Animals grow bigger in water

Animals living in water can grow to a much bigger size than animals on land. The biggest animals in the world live in water. This is because the water supports their mass. The blue whale may have a mass of 143 tons. This is nearly twenty-five times the mass of an African elephant, which is the biggest land animal. Above a certain size, land animals cannot support themselves. The blue whale would not be able to support itself on land.

What is it like to be small?

We have seen that gravity has little effect on small animals. Surface forces such as wind and rain are much more important. To an animal as small as a fly, air is like water, and water is like syrup. A fly can easily be blown away or damaged by the wind. If a dog gets wet it isn't affected much. If a mouse gets wet, movement becomes very difficult. A wet fly carries an enormous amount of water on its surface. Its mass doubles when wet. It can hardly crawl.

Cool it!

An animal loses heat from its body across its surface. Small animals lose heat more quickly than big ones. Can you think why? Answer on page 152. If you were an animal living in a cold climate, would you like to be big or small?

The African elephant lives in hot conditions. Does its size give it any problems?

Did you know?

The fairy-tale giant could never exist. Imagine a man 17.5 ft tall, 3 times the height of an average man. The giant would be 27 times as heavy as the man but only 9 times as strong. If he were shaped like a man he wouldn't be able to walk or even stand up. Jack in "Jack and the Beanstalk" had nothing to worry about!

More about floating in air (pp.88, 122); shapes (pp.24, 25); surface area (pp.25, 129)

The importance of light

The importance of plants

Animals cannot live without plants. Either they eat plants, or they eat animals that eat plants. This is because animals cannot make their own body materials. Think of an animal as being made up of building blocks. The animal cannot make the building blocks, but plants can. The animal can only rearrange the building blocks to make different structures.

Photosynthesis

Green plants use a process called photosynthesis to make their own body materials from simple substances, which they take in from their surroundings. They use carbon dioxide gas from the air and water from the soil. They also use the mineral salts dissolved in water. Energy is needed to join all these simple substances together to make a complicated structure like a plant. This energy comes from sunlight.

▲ A plant is made up of many different raw materials.

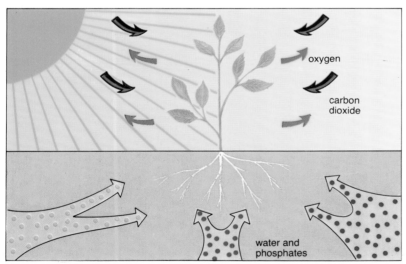

▲ This diagram shows how raw materials enter the plant. Carbon dioxide gas enters through tiny holes in the leaves, called stomata. Water and mineral salts are taken up from the soil by the roots.

Energy from light

Plants contain a green substance called chlorophyll which captures energy from light. Leaves contain a lot of chlorophyll. They are broad and flat to catch as much light as possible. In the leaves the energy from light is used to make all the materials needed by the plant body. During this process the plant produces oxygen gas. This goes off into the atmosphere, and is used by animals for breathing.

Two-way transport systems

The materials made in the leaves are carried to the rest of the plant in a series of tiny tubes. Some of these materials are used to make new plant body — they help the plant to grow. Some are stored in the roots or in special structures such as bulbs and tubers. They can be used for growth later. Another set of tiny tubes carries water and dissolved mineral salts from the roots to the leaves. These two sets of tiny tubes make up the veins of the leaf.

Sun seekers

Plants on a windowsill bend toward the light so their leaves trap light all around the shoots, instead of only on one side.

Catching sunbeams

Plants arrange their leaves so as to catch as much light as possible. If you lie on your back and look up at the leaves on a tree, you will see that hardly any of them overlap. Their stalks twist so that they do not shade each other.

Making animals from plants

You can think of living organisms as being made up of different colored building blocks. The building blocks are really chemicals. Each color represents a different chemical. Instead of cement, living building blocks are held together by energy. The plant acts rather like a bricklayer — it traps energy from sunlight and uses it to make complex chemicals from simpler ones.

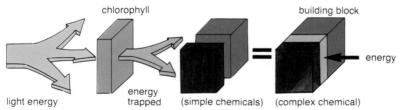

chlorophyll

building block

energy

light energy

energy trapped

(simple chemicals)

(complex chemical)

Animals cannot do this. Animals that feed on plants rearrange these building blocks to make the chemicals they need for their own bodies. Other animals get their building blocks by eating animals that feed on plants.

Animals can also break down the complex chemicals to get the energy out again. They use the energy for moving about and for keeping warm. So all the energy animals use was originally captured from sunlight by green plants. Animals are solar powered!

Leaves have holes

Carbon dioxide gas enters the leaves through hundreds of tiny holes called stomata. Oxygen gas produced by photosynthesis goes out to the plant through these holes. Water also evaporates from the plant through the stomata, so the plant has to take up a lot of water through its roots to avoid wilting.

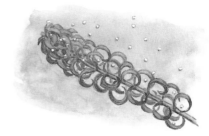

▲ Plants that live in water take in their raw materials from the water. They absorb dissolved gases and mineral salts all over their surfaces. This pondweed is giving off bubbles of oxygen as a result of photosynthesis.

— Did you know? —

Every year plants trap nearly 28 trillion tons of carbon from carbon dioxide gas and turn it into living material.

Most of the oxygen in the atmosphere has been produced by plants.

More about energy in plants and animals (pp.30, 31); animals that eat plants (pp.32, 33)

Energy and life

Living things

If you look at any habitat you will find many different animals and plants. There are probably more than you might think. This is because lots of living organisms are small and are not easily seen. They hide under stones and tree bark, in the soil, on other animals and plants and even inside them.

Why do living things need energy?

All living things carry out important life processes. For example, they move to find new food or to escape danger. They feed. They sense what is going on in their surroundings. They get rid of waste materials. They breathe, and they grow and reproduce. You probably think that plants do not do some of these things, but they do. They don't move as much as animals, but they do all the other things. In order to carry out all these processes, animals and plants need energy.

Capturing energy

There are different forms of energy. Each form can be changed to another form. The Earth receives all its energy from the Sun. There is energy in sunlight. Animals can't do much with this energy, but green plants can. They capture some of the Sun's energy and use it to carry out all their life processes. This capturing of the Sun's energy by green plants is called photosynthesis. It is a very complicated process and you can read more about it on pages 28 and 29. Green plants use the Sun's energy to make their own food. For photosynthesis, they also need carbon dioxide and water. By photosynthesis, green plants make foods that are very rich in energy. They store these inside them ready for use. Animals can also use these foods.

Energy for animals

toucan

orangutan

python

Animals get their energy by eating green plants or by eating other animals. Animals which eat plants are called herbivores. Those which eat other animals are called carnivores. Some animals eat both plants and animals and we call them omnivores. Look at the pictures of these eight animals. Can you group them based on their feeding habits? You will have to think carefully about some of them. All the answers are to be found somewhere in this book. If you get tired of looking, you will find the answers on page 152.

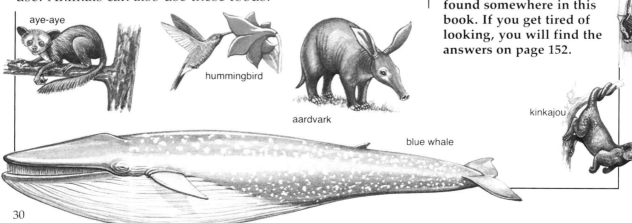

aye-aye

hummingbird

aardvark

blue whale

kinkajou

Food chains

When we think about feeding, we can imagine each plant or animal as a link in a chain. For example, a leaf can be eaten by a caterpillar, and a caterpillar can be eaten by a shrew. This is a three-link food chain. Energy is passed from one link to the other, but it moves in one direction only. Follow the arrows.

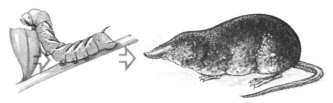

Any habitat has many three-link food chains, though they may be made up of different animals and plants. It is easy to make a four-link chain. All you have to do is add another carnivore on the right. You need an animal which eats shrews.

Why can't you turn a three-link chain into a four-link chain by adding something to the left-hand side? Try making up other food chains. Can you work out a simple food chain from an African habitat? Look at pages 130 and 131 and build a three-link chain from the information given about the wildlife in a South American rain forest.

Losing energy

Some energy is lost as it passes from one link to another along a food chain. It is a slow process. The leaf uses some of the energy captured from the Sun to carry out its life processes. It also wastes some. Therefore, there is less energy for the caterpillar. The caterpillar, in turn, uses some of the energy it got from the leaf and it also wastes some. Thus, there is even less energy to pass on to the shrew. By the time the shrew has used and wasted some of the energy it got from the caterpillar there isn't much left for the snake. Can you now see why most food chains have less than six links?

Many antelopes but few lions

There is a lot of energy at the beginning of a food chain, but not much at the end. The energy at the beginning can support many herbivores. After three or four transfers, there is only enough energy to support a few carnivores. Can you now explain why Africa has lots of antelopes but fewer lions?

Food webs

Many animals eat more than one kind of food. This means they can take food from other food chains. Imagine a food chain made up of these links: leaf → slug → toad. A lot of animals eat leaves. There are also animals which eat slugs and toads. These animals form new food chains. The food chains connect with each other to form food webs. Can you make up a food web using the animals and plants in your garden?

Remember

There is no shortage of energy from the Sun. It reaches the Earth every day. But raw materials like carbon dioxide, water and minerals are limited. They have to be recycled. The decomposers do this.

More about decomposers (pp.40, 41); how plants capture sunlight (pp.28, 29); how different animals feed (pp.32-39)

The vegetarians

▲ Bison grazing on the North American plains.

A tough diet

Unlike plants, animals cannot make their own food. Instead, they have to feed on plants or other animals. Animals which feed mainly on plants are called herbivores.

Plants contain a lot of tough fibers. Many vegetarian insects, like locusts and caterpillars, have hard horny mouthparts which help them cut up leaves. Other vegetarian animals like tortoises and turtles use sharp horny jaws, while slugs and snails use a rasping tongue covered in tiny teeth. Some birds have special bills — finches have large heavy beaks for cracking open seeds, and parrots have hooked bills for tearing into fruits.

Plant-eating mammals have large flat-topped teeth with hard ridges for grinding leaves to break down the plant fibers. The grinding motion wears the teeth down very quickly. Because of this, the teeth grow all the time. Cows and sheep move their jaws sideways as they chew, which helps them grind the food. Try moving your jaws sideways — can you hear your teeth grinding together?

Reach for it

Some animals are specially adapted for eating certain kinds of plant food. Rodents and monkeys have flexible fingers for grasping and picking at food. The giraffe has a long neck which allows it to feed on leaves that other animals cannot reach.

▲ Reach for it! The elephant uses its trunk to pull down branches.

Eating all the time

Plant cells have thick walls made of a material called cellulose, which is very difficult to digest. The plant leaves have to be chewed for a very long time. Some animals, like cows, sheep and deer, have special stomachs for storing undigested food, so they can take in a lot of food in a short time and chew it later. When they have finished feeding, they find a safe sheltered place, and bring the food back into their mouths to chew it. This is called chewing the cud. Cows sit for hours just chewing grass.

A secret army

Animals like cows have a secret army of helpers. In their stomachs are special bacteria which are good at breaking down cellulose. The cow supplies the bacteria with food, and the bacteria help the cow digest the grass.

Liquid food

A few animals feed on nectar from flowers. Butterflies and moths have long tubelike mouthparts for sucking up the nectar. The mouthparts reach deep inside the flower, and can be rolled up out of the way when not in use. Many flowers have special lines and patterns which guide insects to the nectar. Hummingbirds use brushlike tongues to lap up nectar, and so do some bats. Bees not only feed on nectar, but they also collect pollen to eat later, storing it in bristly baskets on their legs.

▲ Squirrels, mice and chipmunks have long, sharp front teeth for breaking open nuts, and flexible fingers for holding them.

Did you know?

An adult elephant needs to eat 300lb of plants a day.

Beavers' teeth are so strong that they can fell trees by gnawing through the tough wood.

Leaf-cutter ants grow their own food. They cultivate fungi in special underground gardens.

More about how plants make their own food (pp.28, 29); leaf-cutter ants (p.44); hummingbirds (pp.74, 91)

Filter feeders

Food you cannot see

Floating in the water of lakes, ponds, rivers and the oceans are millions of tiny plants and animals which cannot be seen by the naked eye. There are also many small particles of dead organic material. Some of them are excreted by living organisms, others are the broken-up remains of dead plants and animals.

Many different animals, both large and small, feed on these tiny pieces of food. They are filter feeders, trapping their food by passing the water through a fine net or sieve. This lets the water through but holds back the particles.

Bristly legs

When they are feeding, barnacles open their cases and comb the water with their bristly legs. Some shrimps and water fleas also use bristle-fringed legs to filter food from the water.

▲ Barnacles filter feeding.

Moving water

Some animals use rows of tiny beating hairs to make their own water currents to bring food in. Sponges are filter feeders — if you look closely at the surface of a sponge you will see lots of tiny holes through which water enters and leaves. Inside the sponge, food particles in the water are trapped by tiny hairs.

Mussels, oysters and clams draw water over their gills where tiny hairs filter out food particles.

Herring and their relatives also use specially adapted gills to sieve food from the water. They swim along with their mouths open so that water flows over the gills.

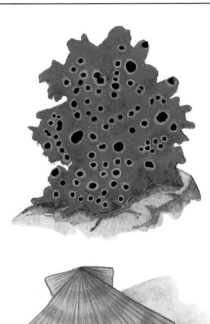

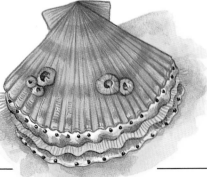

▲ Fan worms live in tubes or burrows on the seabed, and put out a fan of feathery tentacles which sift the water for food. Tiny hairs on the tentacles sweep the food particles onto a moving sheet of sticky mucus which carries them into the worm's mouth.

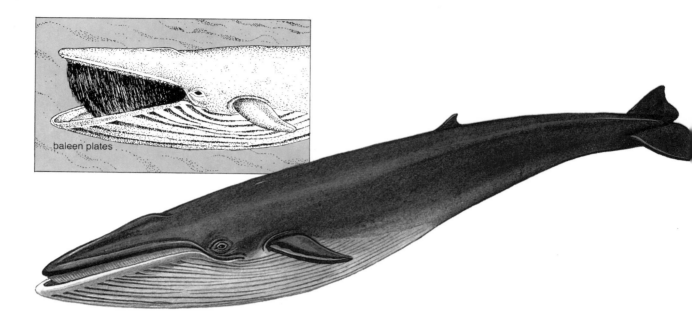

baleen plates

Giant filter feeders

Baleen whales have no teeth. Instead, they have about four hundred large plates of whalebone or "baleen," which act as a strainer. The whale takes in enormous mouthfuls of water. It then forces the water out over the baleen plates and licks off the food that is left behind, mostly small creatures like plankton and shrimps. The mouth of a baleen whale may take up almost one-third of its body.

▲ Flamingos have strange upside-down beaks which they sweep through the water. The edges of the bills are lined with rows of horny plates, forming a strainer that lets water through but traps tiny organisms.

-Did you know?-

The largest animal in the world, the blue whale, is a filter feeder. It is about 82 ft long, and weighs about 143 tons, about the size of four large dinosaurs. It will eat 11 tons of shrimps in one meal. Each gulp filters 58 quarts of water.

The largest bivalve in the world, the giant clam, is also a filter feeder. It can weigh up to 1,000 lb.

Water fleas have such fine filters that they can trap bacteria which are only 1/25,000 in long.

More about filter feeders in rock pools (p.113); filter-feeding fish (p.117)

35

Carnivores — the meat-eaters

Carnivores

Animals which eat other animals are called carnivores, which means meat-eaters. Unlike plants, animals move, so carnivores first have to catch their prey.

Ambush!

Many animals ambush their prey. The praying mantis stays very still, its color blending with its background. When an insect comes within reach, it grabs it with its long front legs. In the lake, pike lie in wait for smaller fish, their striped bodies matching the pattern of sunlight shining through the water weeds. Crocodiles and alligators can lurk just below the water surface, with only their eyes and nostrils above the water, waiting for animals to come to drink at the water's edge.

The chase

Other animals stalk their prey, relying on a fast pounce or a quick chase to catch it. White pelicans work in a team, forming a circle around a school of fish and driving them into the center. Lions also hunt in groups to cut off individual animals from a herd, while cheetahs rely on outrunning their prey. Army ants use sheer weight of numbers. Hundreds of them swarm over any small creature on the forest floor, biting and stinging it until it dies.

Stingers and biters

Some animals can take quite large prey by biting or stinging it. Snakes may strike at an animal with poison fangs, then track it until it dies. When they have trapped a fly, spiders rush in and bite it, waiting for their paralyzing poison to take effect before wrapping up the victim. Sea anemones and jellyfish have stinging tentacles which shoot poisoned hooks into any small animal that touches them, paralyzing and trapping it at the same time.

▲ Portuguese man-o'-war jellyfish with trapped fish.

Clingers

Octopuses, squids and cuttlefish have long curling tentacles armed with rows of suckers which cling to the prey and sweep it into the mouth. Constrictor snakes coil around their victims and squeeze them until they suffocate to death.

Biting and sucking

Meat-eating mammals have special teeth for dealing with their prey. In the front of the mouth are sharp incisors for cutting into meat and scraping it off the bone. Behind them is a pair of long, curved, pointed canines for piercing and stabbing the prey and holding it as it struggles. Next are the big carnassial teeth, which act like shears to slice off flesh and crack bones. Finally, there are the flatter molars for crushing bones and flesh.

Spiders handle their prey the easy way. They wrap it safely in bands of silk and dribble digestive juices onto it. They then suck up the dissolved remains.

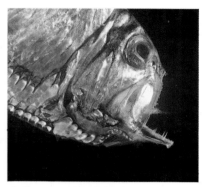

▲ Some animals just open their mouths and wait for their prey to fall in. Swallows, swifts and whippoorwills fly along with their mouths open to catch flying insects, while the deep-sea hatchetfish swims with its huge mouth wide open to catch fish it can hardly see in the darkness.

▲ An egg-eating snake swallowing the egg of a small bird.

▲ The cheetah's teeth are adapted for eating meat.

Swallowing specials

Some animals swallow their prey whole. The master swallowers are the snakes. They can actually unhinge their jaws to swallow large prey, and their skin is very elastic, stretching around the prey as it moves back through the snake's body.

Stealing a living

Not all carnivores catch their own prey. Some steal the prey of others. Skuas are seabirds which attack other seabirds in the air until they drop the fish they have just caught. Bears will drive wolves off their prey, and packs of hyenas will drive other African mammals, such as lions, away from the carcass they have just killed.

—— *Did you know?* ——

The cheetah is the fastest animal on land. It can run at speeds of 70 mi/h while chasing its prey.

The spine-tailed swift, from Asia, moves even faster, reaching a speed of 105 mi/h when flying.

A column of army ants is like a super-animal weighing over 440 lb, with 20 million mouths and stings.

More about animal trappers (pp. 48, 49); deceits and disguises (pp. 52, 53); hunting in the dark (pp. 118, 119, 126, 127)

The insect-eaters

▲ The insect-eating bats hunt in darkness. They find their prey by "echolocation." They utter high-pitched squeaks and listen for the echo to bounce back from the flying insects. Their teeth are similar to those of other insectivores, but they use their wing tips to flick insects into their mouths.

The star-nosed mole

The star-nosed mole has a remarkable fleshy star on its nose which is loaded with 22 sensitive feelers which it uses to help find its food.

Insectivores

Animals which feed mainly on insects are usually specialists. Some have special teeth for crushing the hard shells of insects, others have special ways of catching hundreds of tiny insects at a time. Some catch insects in the air, others hunt them underground.

One group of mammals specializes in eating insects. They are called the insectivores. They include shrews, hedgehogs and moles. The teeth of an insectivore look very alike, with pointed ends for seizing and crushing their prey.

Most insectivores rely on sound and smell to find their prey. Many, like the shrew, have flexible snouts with long sensitive whiskers. Insectivores are usually nocturnal, but some tiny shrews hunt by day and by night. Moles live in the dark most of the time, tunnelling through the soil with huge scoop-shaped paws in search of insects and worms.

Close your nose

In the African grasslands the aardvark, or earth pig, raids termite mounds, using its strong claws and long sticky tongue. Its piglike snout has a fringe of bristles to keep off the termites, and it can close its nostrils for extra protection.

Sticky tongues

Some animals specialize in eating only ants and their relatives, the termites. These are such small insects that they have to eat a lot of them. The anteater is a big lumbering animal with a very good sense of smell and a very long snout. It digs termites out of their large earth mounds with its huge curved claws. Once it has made a hole in the termite mound, it pushes in its long sticky tongue. As many as five hundred termites at a time stick to the tongue and are then swallowed. The anteater has a thick skin which protects it against the bites of the termites. Anteaters have no teeth, but grind up the insects using their powerful stomach muscles.

▲ The anteater uses its sticky tongue to pick up its prey. Other insect-eaters also use sticky tongues. The woodpecker drills into tree trunks to find insect grubs, then uses its tongue to get them out. Frogs and toads flick out their tongues to catch passing flies.

The most amazing tongue is that of the chameleon. When stretched out, it is about one foot long — twice its body length. When not in use, the tongue is folded like a concertina in the mouth. The chameleon can change color to match its background, so that it is almost invisible to its prey. Its eyes are on little turrets and can look in different directions at the same time, useful for spotting approaching insects. Once it has seen its prey, the chameleon focuses both eyes on it and very slowly gets into position to attack, holding on to its branch firmly. Then the tongue flicks out, the insect sticks to its tip, and it flicks back into the mouth, all in the space of four-hundredths of a second.

Did you know?

Savi's white-toothed pigmy shrew, an insect-eater, is the smallest mammal in the world, only 2.3 in long, including its tail. It weighs 0.05 oz.

The giant anteater has a tongue 23 in long.

A toad can flick its tongue out and back in 1/10th of a second.

The aardvark is one of the fastest burrowers in the world. It can dig a hole with its claws faster than a man can with a spade.

More about ants and anteaters (p.105); animal senses (pp.50, 51)

Scavengers and decomposers

▼ Small corpses are often buried. Burying beetles dig away the soil underneath the corpse to form a pit, into which the corpse falls. Then they lay their eggs in the dead flesh. The eggs hatch into grubs which feed on the corpse.

The great disappearing act

Plants and animals are always multiplying. If they lived forever, the Earth would soon become overcrowded. But plants and animals die of old age or disease, or are killed and eaten by other animals. This prevents their numbers from becoming too large.

Yet we do not see many dead animals and plants. The leaves that fall in autumn slowly disappear, old tree trunks rot, and animal carcasses are soon reduced to heaps of bones. Many different organisms are involved in this disappearing act. Scavengers feed off the carcasses, and decomposers break down the dead organic matter into smaller and smaller pieces. The mineral salts and other nutrients of the dead material are released into the soil or the sea, ready to become part of new living organisms.

Scavengers large and small

Some animals specialize in scavenging. Hyenas and jackals of the African plains can hunt their own prey, but usually feed on animals which are already dead — the remains of lion kills, or animals that have died of disease or old age.

Vultures eat only carrion (dead meat). They are big, powerful birds with hooked bills for stripping skin and flesh. Their heads and necks are usually bald, with no feathers to be messed up when they are feeding on the carcasses.

Along the seashore, seagulls scavenge for food washed ashore by the tide. They also follow ships at sea to feed on the waste that is thrown overboard.

▼ Dung beetles bury animal dung, making it into a large ball which they roll to a safe place.

Underwater scavengers

In the sea, some fish scavenge on the corpses of other fish. The hagfish has a long eel-like body which it wriggles into carcasses. Mullet feed on particles of organic matter which sink to the seabed, feeling for them with tentacles on their snouts. Crabs scavenge along the shore and on the seabed.

▲ Crab scavenging on the seashore.

Many different sea creatures, from shrimps and barnacles to sea fans and tube worms, filter organic particles from the water. Some tiny pond animals feed like this too.

If it were not for all the scavengers and decomposers on land and in the sea, the living world would soon run out of nutrients, and there would be no future generations of living creatures.

◀ The most important decomposers are the fungi and bacteria. The fungi form a hidden network of threads covering the undersides of dead leaves, rotting wood, dung, carcasses and decaying pieces of organic matter in the soil. These fungus threads pour digestive juices on the dead material and absorb it. Bacteria feed on dead material, helping it to decay. Millions of them occur in the soil, in ponds and lakes, and in the sea.

Hidden decomposers

Flies lay their eggs in corpses and dung pats. The eggs hatch into maggots which feed on the dead meat or the dung. Some moth larvae can even digest the horns of dead animals. In the tropics, termites feed on wood, and all over the world wood-boring beetles, like the death-watch beetles, and their grubs, the "woodworms," chew tunnels in tree trunks, furniture and house timbers.

Slugs and snails, millipedes, woodlice and worms all feed on dead and decaying organic matter in the soil.

Did you know?

A pair of burying beetles can bury a mouse in just a few minutes.

Museums use carrion beetles to clean bones for exhibits.

A dung beetle only 0.8 in long can bury 6.1 in^3 of dung in one night.

More about life in the soil (p.94); scavengers on the seashore (p.113)

Living together

Feeding each other

Some of the commonest partners in symbiosis are algae. These are tiny one-celled plants, small enough to live inside the cells of other organisms. Algae make their own food by a process called photosynthesis. They use water, carbon dioxide gas from the air, and energy from sunlight. During this process they produce oxygen, which animals need to breathe, and sugars and proteins, which are then used by their partners. Animals breathe out carbon dioxide, the gas which the algae need for photosynthesis.

Many algae live protected inside animal cells. The bright colors of corals, the green of hydra and the brilliant blues of the mantles of giant clams (below) are all caused by algae in their tissues.

Symbiosis

Sometimes quite different animals and plants live and work together. This kind of close relationship between different living organisms is called symbiosis, which means "living together."

One example of symbiosis can be seen in Africa. A small bird, called a honeyguide, likes to eat wild bee grubs and wax, but the honeyguide cannot break open the bees' nest to reach them. So it attracts the attention of a honey badger (who is fond of honey) by bobbing in front of it and leading it to the bees' nest. The badger uses its sharp claws to tear open the nest, so that both of them can feed. In this case both partners are better off together than on their own. This kind of symbiosis is called mutualism.

The protection game

Some symbiotic animals can get on very well without their partners, but they usually choose to live or work together. Hermit crabs often have sea anemones riding on their shells. Some hermit crabs will actually pick up the anemones and place them on their backs. It seems likely that the anemone's stinging tentacles put off predators who might attack the crab, while the anemone gets a free ride to new feeding grounds. It also eats the crab's leftovers.

Nature's cleaners

Another protection game occurs in the sea. This time the enemies are parasites — animals that live on, or inside, other animals or plants, and get food or shelter from them. In the sea, some parasites cling to the scales of fish, and feed on their flesh. In turn, cleaner fish feed on these parasites. This cleaning activity is very important to the fish — so important that fish that are usually enemies queue together at special "cleaning stations." Many fish spend as much time being cleaned as they do hunting for food. The cleaner has easy-to-recognize bright stripes. To signal that it is a cleaner and not a tasty meal it does a special dance. The large fish will open their mouths, and even their gills, for the little cleaner.

On land, tick-birds and oxpeckers stand on cattle and other grazing animals, and feed on the parasites in their fur and skin. One bird, the Egyptian plover, is said to clean inside the open mouths of crocodiles.

▲ A lichen is made up of a fungus and an alga, which work together as one organism. Lichens can grow even on bare rock. They are found all over the world, from the hottest deserts to the edges of permanent snow and ice where no other plants can grow.

Mutual protection

Brightly colored clownfish live among the stinging tentacles of coral reef anemones. These protect them from enemies. But the clownfish also protect the anemone, chasing off other fish that would like to eat the soft tentacles.

—Did you know?—

The smallest zoo on Earth? A termite just a tenth of an inch or so long is not a single organism, but anything from 10 to 100 different organisms, all depending upon each other. In addition, bacteria and other tiny one-celled organisms live in its gut. They break down the tough fibers of the wood it eats, and make substances the termite can use.

Many plants, especially forest trees, have a layer of fungi around their roots. These fungi absorb valuable nutrients from the soil and pass them to the plant.

More about living together (pp.44, 45); corals (pp.120,121); photosynthesis (pp.28, 29); parasites (p.45)

Living together

One-sided friendships

It is easy to see how both partners can benefit from living together. But sometimes only one partner appears to benefit. The antbirds of tropical forests use ants to flush out their prey. They fly just ahead of the marching columns of army ants, feeding on the insects fleeing before them. But the ants do not benefit. This type of association is called commensalism, which means "eating at the same table." The partner which benefits is called the commensal.

▼ Humans are also involved in symbiosis. In our guts we have millions of bacteria, which help us break down the vegetables and fruit we eat. We farm crops and animals for food and milk. We also grow fungi — mushrooms to eat, yeast to make bread and alcohol, and microscopic fungi and bacteria to make cheese and yogurt.

Fungus gardens

Ants form many symbiotic partnerships. Leaf-cutter ants cut up vast numbers of leaves and take them to a garden deep inside the ant nest. Here they grow a special fungus, which is found only in ant nests. The ants cannot digest the tough plant food, but the fungus can. It grows special knobs of fungus for the ants to eat. The ants provide the fungus with leaf pieces and a warm shelter. They also weed out any other fungi that try to grow in the garden.

▲ Leaf-cutter ants on trail.　　　▼ Ants with aphids.

Ant farmers

Other ants farm aphids (greenfly), sheltering them in the ant nest at night and leading them up the food plants to eat by day, guarding them against predators. The aphids feed on plant sap, but they take in too much sugar. This extra sugar is squeezed out of the aphids as droplets of honeydew. The ants can make the aphids produce honeydew by stroking them. Then the ants feed on the honeydew.

Unwelcome guests

Not all partners are welcome. Many animals and plants live at the expense of others. They are called parasites. Fleas and lice cling to the skin of other animals (their "hosts") and suck their blood. Mosquitoes are temporary visitors, stopping only for a quick drink of blood. Leeches, ticks and mites have bodies like elastic bags, which stretch as they fill up with blood. Some ticks can live for a whole year on a single meal of blood.

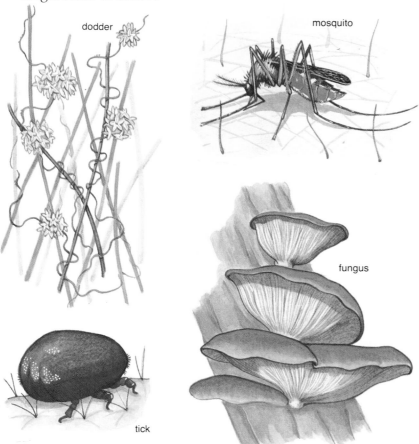

dodder

mosquito

fungus

tick

▲ Flowering plants can also be parasites. Mistletoe grows on the branches of trees. Dodder twines up the stems of other plants, putting suckers into them to take up their sap.

Some parasites live inside their hosts. Tapeworms live in the guts of other animals, absorbing the ready-digested food and enjoying warmth and protection. Many fungi live in, or on, other living organisms. Some cause lung and skin diseases, including athlete's foot. Others live in the trunks of living trees, making them rot.

Animal, vegetable or mineral?

This coral is really all three. The coral is an animal which secretes a mineral skeleton around itself, and contains in its tissues colorful tiny plants (algae).

-- *Did you know?* --

The bull's horn acacia tree has its own private army — colonies of ants living in its hollow thorns. These ants drive off any animal, large or small, that tries to feed on the acacia. They even weed out stray seedlings from around the base of the tree. As well as shelter, the tree provides food for the ants in the form of sugary nectar and special round knobs containing oil and protein.

The largest flower in the world, *Rafflesia*, is a parasite of tropical vines. The flower is up to 39 in wide, and weighs 15.5 lb.

More about living together (pp.42, 43); corals (pp.120, 121); plant-eating bacteria (p.43)

Plants of prey

Fact or fiction

A famous author, H. G. Wells, once wrote a story called *The First Men in the Moon*. It told of some plants which could capture and eat humans. Another book, by John Wyndham, called *The Day of the Triffids*, told the story of a group of plants which tried to destroy all humans. Neither of these stories was based on fact. Plants like that do not exist. However, some plants can trap very small animals, such as insects. The animal is killed, often by drowning, or simply by being digested by plant juices as it lies trapped. Sometimes the plants catch even larger animals, like baby frogs. Plants which trap animals are called carnivorous plants.

Plants and energy

All green plants make their own food. They capture the Sun's energy and use it to make sugars in their leaves. This process is called photosynthesis. Green plants also take in mineral salts from the soil. They use these salts and the sugars they make to build all the materials they need. Some plants live in poor soils which do not have all the important mineral salts. Boggy soils and marshland are like this. The plants living in them must find other sources of minerals. Carnivorous plants are sometimes found in marshy and boggy areas, although they also grow in other soils, especially in the tropics.

Different types of trap

▼ **Sticky-surface trap** (sundew)
This trap captures its prey by means of a sticky substance. Other traps have small tentacles on the surface of the leaf. These have sticky heads which catch insects.

▼ **Snap-trap** (Venus fly-trap)
This is made of a hinged leaf. When the trap is set the leaf lies flat and open. An insect landing on the leaf touches sensitive hairs. These trigger the leaf to close and the insect is trapped inside.

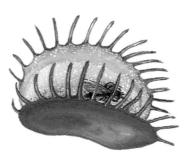

▲ **Jug-of-water trap**
(pitcher plant)
The leaf is shaped like a jug or pitcher. Insects landing on the rim of the jug slide down the slippery slope to the inside. They land at the bottom of the jug. Waxy scales make the slope slippery. They stick to the insect's feet like small skis, and make it slide down quickly.

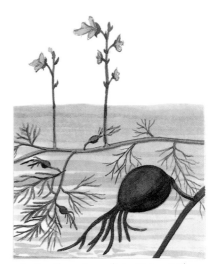

▲ **Suction trap** (bladderwort)
Suction traps are usually found in water plants. Each trap is like a tiny ball. It has a small trapdoor or lid. This opens to let the animal be sucked in with a rush of water. The trapdoor closes once the animal is caught.

Trapping the prey

There are many different kinds of carnivorous plant. They are found in various shapes and sizes. Even though they look different, these plants have some things in common. They have a bait to attract their prey. They also have a trap in which to catch the prey. Plants are not able to move like animals. Instead, they have to lure or attract their prey to them. They do this by means of color, smell and a sugary liquid called nectar. If the animal takes the bait it becomes trapped. After this it is killed and digested. In some plants, the trap must be reset after the animal has been eaten. The plant can then catch another animal. Carnivorous plants have different kinds of trap and they catch their prey in different ways. The pictures opposite show the four main types of trap used by carnivorous plants.

The body-snatcher

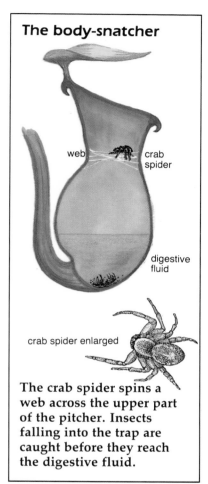

The crab spider spins a web across the upper part of the pitcher. Insects falling into the trap are caught before they reach the digestive fluid.

The daredevils inside traps

Some animals are able to live quite happily within these deadly traps. A small bug lives on the leaves of a plant called *Byblis* which grows in Australia. This plant sets a sticky-surface trap like the ones set by sundews. The small bug is able to walk on the trap without getting stuck. It avoids capture and digestion. We don't yet know how this bug escapes being caught and eaten. A species of fly has found a way of living inside the trap of the American pitcher plant, *Sarracenia*. The maggots of the fly crawl in the digestive juices at the bottom of the trap, but they are not digested. A species of mosquito also lives in the same sort of trap. It is even able to carry out its life cycle within the trap. The adult mosquito enters and leaves the trap like a tiny helicopter. Even small bladderwort traps contain animals living inside them. It is puzzling that the animals do not come to any harm.

Did you know?

Some pitcher plants may catch thousands of insects in a few weeks.

The Portuguese sundew takes about a day to digest a small insect.

A Venus fly-trap can catch a small frog.

More about photosynthesis (pp.28, 29); trappers (pp.48, 49)

Animal trappers

Come into my parlor

Most animals find their food by moving around. Plant eaters need to move around to find new vegetation on which to feed. Carnivores rely on their senses and speed to catch their prey. Some animals have solved the problem of catching food in another way. These animals live a much less active life. Instead of looking for food, they let their prey come to them. These are the "trappers" of the animal world.

Different kinds of trap

There are many different kinds of trap. Each one is designed to catch its victim in a different way. Some traps are simple in structure and need few building skills. Others are very complicated and take a long time to make. Most "trappers" catch their prey by means of pits, trapdoors, snares, nets or webs. Some specialized animals use a "fishing line" technique. Whatever type of trap is used, it has to be kept in good working order. The animal must also repair the trap if it becomes damaged. Some traps must be reset when they have been used.

A silken trap

A poisonous spider from Australia builds a trap similar to that of the ant lion. It makes a burrow lined with silk. The entrance to the burrow is woven into a funnel shape. The spider feeds on small animals which fall into the funnel.

The waiting jaws

The ant lion, found mainly in the tropics, is the larval form of an insect which looks like a damsel fly. The female lays her eggs in fine, sandy soil in a sunny place. When each egg hatches, the tiny ant lion larva digs a funnel-shaped pit. The ant lion hides at the bottom of the pit with only its head and pincerlike jaws showing. If an insect, such as an ant, falls into the pit, it slides to the bottom and into the large, powerful jaws of the ant lion.

Behind the trapdoor

Trapdoor spiders hunt by night. The trapdoor is kept closed during the day. At night, the spider keeps the door half open, its front legs sticking out. When an insect comes near it touches trip lines of silk which the spider has laid down. This alerts the spider which pounces on the insect and drags it into its tube. The trapdoor closes automatically.

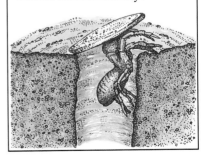

Setting a trap

1. The garden spider sets its trap by making a web of tiny threads made of silk. In the spider's abdomen silk glands release liquid silk. The silk hardens when it reaches the air. The spider begins the web by casting a thread into the breeze. The thread floats in the air until it catches on a solid object like a twig.

2. The thread links two twigs. The spider walks up and down the thread making it bigger and stronger by adding more threads.

3. By climbing up and dropping down the spider builds a framework.

4. Then the spider spins the "spokes" of the web.

5. A temporary thread is spun from the center of the web, in a spiral, to the outside. This strengthens the web.

6. Finally, the spider covers the web with a thread coated with tiny blobs of sticky gum. Starting from the outside, it removes the temporary thread. In its place it puts the gummy thread.

7. Flies and other insects that fly into the web stick to the thread, unable to escape. The spider itself is able to move around the web without sticking to it. This is because it walks only on the nonsticky spokes between the spirals.

The larva of one kind of caddis fly lives in water. It builds a "fishing net" from silk. The net is funnel-shaped. The larva stays in the narrow part of the funnel. It feeds on small animals caught in the net. The flowing water keeps the net open.

—Did you know?—

An Australian spider called *Dinopis* holds a sheet of sticky web in its front legs. It drops the sheet over a passing insect, like a gladiator's net.

The bolas spider, from Australia, dangles a long thread with a sticky blob on its end. It swings the thread round and round and catches insects on the sticky blob.

Glowworms living in caves in New Zealand catch their prey on fine threads hung from the roof. The threads are covered with glowing droplets to attract the prey.

More about carnivores (pp.36, 37); animals in ambush (pp.109, 111); animal trappers (p.119)

Animal senses

Senses

We use our senses to find out about the world around us. We look, listen, smell, taste and touch. Most of all we look. Animals have the same senses, but different animals rely on different senses as their main source of information. Animals use their senses to find food, to warn them of danger, and even to attract a mate.

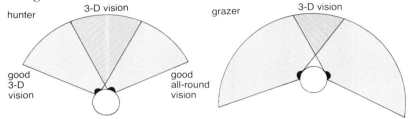

Seeing is believing

You can tell a lot about an animal's way of life by looking at its eyes. If it relies a lot on sight, its eyes will be relatively big. If it is a hunting animal, like the tiger, its eyes will be placed toward the front of its head, so that the fields of view of the two eyes overlap. This allows it to judge distance accurately for pouncing on prey. Animals with many predators, like the rabbit, usually have eyes at the sides of their heads. They can spot a predator coming from almost every angle, but they are not very good at judging distance.

Not all animals see in color. Many see only in black-and-white. Some, like dogs and squirrels, are partly color-blind, and cannot see reds and greens. Their world is a blue and yellow one. Bees can see colors we cannot see. They see ultraviolet light. Many flowers look dull to us, but have bright patterns on them when viewed by a bee.

▼ Insects have compound eyes made up of lots of tiny lenses.

Some unusual senses

Some animals have senses we do not have. The rattlesnake has hollows on each side of its head which contain heat-sensing cells. It can find warm-blooded prey, like mice, even in the dark, by sensing the direction from which the warmth is coming. Some fish can give off and detect electricity. They use electricity to find their prey, rather like bats use echolocation. Most animals can detect gravity, and use the information to keep themselves the right way up, and to balance. Some birds and whales can even detect the Earth's magnetism, and use it for navigation.

Feeling the vibrations

Fish and tadpoles have a sensitive line running along the side of the body which can detect movements in the water.

▼ The chameleon can look in two directions at once.

Smell and taste

Most animals have a much better sense of smell than we do. Backboned animals use their nostrils to smell. Insects have scent detectors on their feelers. Male moths have very large feelers to pick up the special scent given off by the female moth.

We taste with our tongues, but many animals, including fish, have taste sensors all over their bodies. Butterflies taste with their feet.

Hearing without ears

Only mammals have big ear flaps. Some mammals, such as dogs and deer, can twist their ear flaps toward the source of sound. Crocodiles and seals close their ears when they are underwater. Frogs, lizards and birds have no ear flaps at all.

Hearing is particularly important for animals which hunt and feed by night. Bats find their prey by sending out high-pitched sounds (so high that we cannot hear them), and listening for the echoes. This is called echolocation.

▲ The snake's flickering forked tongue wafts scent particles back to special smell sensors in the roof of its mouth.

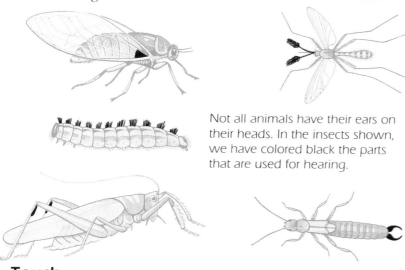

Not all animals have their ears on their heads. In the insects shown, we have colored black the parts that are used for hearing.

Touch

Our skin is covered in tiny cells which detect pressure. This gives us the sense of touch. Our fingers are especially sensitive to touch. We use them to handle and investigate objects. Insects have touch sensors on their antennae. Watch a wasp crawling across a path. See how it waves its antennae over the ground ahead. Touch is particularly important for nocturnal animals, such as cats and mice. The long whiskers on their snouts are sensitive to touch.

Did you know?

Fish have no eyelids. They sleep with their eyes open.

The largest eye in the world belongs to the giant squid. It is 12 ft in diameter, 123 times as big as the human eye.

Sound travels faster in water than in air. Whales can hear each other's calls when they are hundreds of miles apart.

The tip of your tongue is more sensitive to touch than your fingertips. Why do you think that is? Answer on page 152.

More about using senses in the dark (pp.118, 125, 126, 127); attracting a mate (pp.64, 65)

Camouflage

The art of disguise

Some animals are very difficult to see. They match their backgrounds, or look like other objects. It can be useful to have such disguises. The animals may need to hide from their enemies, or they may need to ambush or sneak up on their prey without being noticed.

Many animals simply blend in to their backgrounds. Moths that rest on tree bark may be mottled gray and brown. Lions blend with the dried grasses of the African plains, and fish are often silvery blue. Many tiny animals that float in the sunlit waters of lakes and oceans are transparent, so that their background shows through them.

Some animals look like inedible objects. Moths, grasshoppers, katydids, and even tropical toads, may look like dead leaves, with veinlike patterns. Desert mantids often look like stones.

▲ This praying mantis looks like a flower. It catches insects which fly straight into its arms.

▲ This katydid is pretending to be a dead leaf.

▲ How many moths can you see on this tree?

Confusing colors

How many zebra can you see? Predators often recognize their prey by its shape. Patterns like the stripes of a zebra and the blotches of a giraffe break up the animal's outline.

Acting the part

Just matching the background is not always a good enough disguise. If an animal moves, its shadow may give it away. Many camouflaged caterpillars and moths do not move at all by day, but feed at night. Stick insects and leaflike praying mantids will stay motionless by day, or sway gently in the breeze. The potoo is a bird the color of tree bark. When resting, it sits on top of a small tree trunk and flattens its head on its shoulders to look like the end of a branch.

Changing color

Some animals can change color. The chameleon and related lizards are good at this. They can change from pale green to dark gray on different backgrounds and in different lighting conditions. The plaice, which lives on the seabed, can even match a chessboard put under it.

Hiding shadows

Shadows are a real giveaway. In the sunshine, an animal looks paler on top and darker underneath as a result of shading. Some moths have fringes along the edges of their wings to break up their shadows when they are resting on tree trunks.

Mammals like the antelopes that graze on the African plains often have "countershading." They are light below and dark above, so the effect of shading is canceled out.

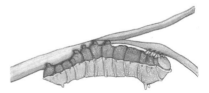

▲ Caterpillars that live upside-down on twigs have their shading the 'wrong' way up.

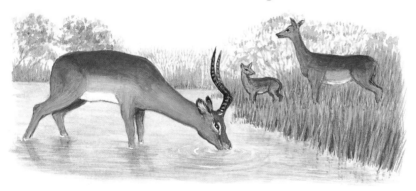

In the sea, fish are often dark on top so that seabirds will not see them against the dark water below, and light underneath, so that predators below them cannot see them against the bright sunlit water above. Some fish have shiny reflective sides so that they do not look like dark shapes when seen from the side.

— Did you know? —

Some caterpillars and spiders look just like bird droppings.

Decorator crabs attach pieces of seaweed and other materials to their backs so that they look like their surroundings.

Some plant bugs are shaped like thorns to deceive their enemies.

More about defense (pp.54, 55); camouflage (pp.102, 117, 118, 121); countershading (p.84)

On the defensive

Food for thought

With so many animals living by killing and eating other animals, it is not surprising that the victims have developed various ways of defending themselves. Plants, too, have defenses against the animals that would like to eat them.

Armor plating

Some animals, like the armadillo, have their own armor plating. Oysters and mussels can simply shut their shells, and snails and turtles can retreat inside their shells. Horns, antlers and claws can all be useful weapons for defending yourself. Many caterpillars are covered in spines which make them unpleasant to eat. Plants also use spines to keep off browsing animals. Who wants to eat a cactus? Or a thorny rose stem?

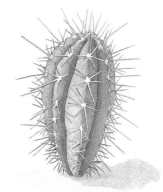

▲ Who wants to eat a cactus?

▼ Sometimes it is enough just to startle the attacker. Many insects have brilliantly colored under-wings which flash into view as they fly off. Some lizards have colored throat pouches that can inflate very quickly. The frilled lizard can unfold a huge fan of skin all around its head, making itself look very fierce indeed. Squids and octopuses squirt out clouds of purple "ink" which hide their escape.

Larger than life

A quick increase in size can scare off a predator. Frightened cats arch their backs, fluff out their fur and hiss. Owls will spread their wings and raise their feathers if disturbed. The puffer is a fish that quickly inflates itself into a round ball covered in spines, making it impossible to swallow.

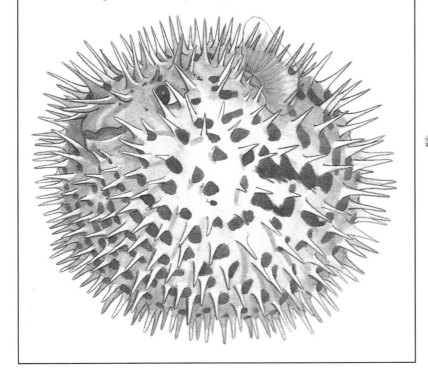

Leave a bit behind

Some lizards have tails which can break off, leaving a surprised predator holding the twitching tail. The lizard grows a new tail later. Many butterflies and moths have a loose coating of scales on their wings, which they can leave behind as they flee.

Stingers and biters

The best defense is sometimes attack. Bees, wasps and hornets have stings. Many animals, from ants to snakes, will bite if attacked. Often they are brightly colored so that a predator will easily recognize them and soon learn to leave them alone.

A nasty taste

Warning colors are also used by animals which taste unpleasant, like some caterpillars, frogs and salamanders. Whip scorpions or "vinegaroons" squirt acid at their attackers, and skunks produce a foul-smelling liquid. Some plants ooze poisonous sap if injured.

Poisonous hairs and spines are also good defenses. It can be fatal to step on the camouflaged spines of a stonefish. Hairy caterpillars can cause skin rashes, and poison ivy and stinging nettles are definitely not for picking.

Running away

You can always try running away. Leaf hoppers and frogs can leap great distances if threatened. Hares and antelopes are excellent long-distance runners, and can soon tire out a predator that gives chase.

Dead or alive?

If all else fails, you can pretend to be dead, as many predators will only attack moving animals. Snakes and opossums and many beetles use this trick, staying perfectly still and limp until the enemy has gone away.

▼ Snake pretending to be dead.

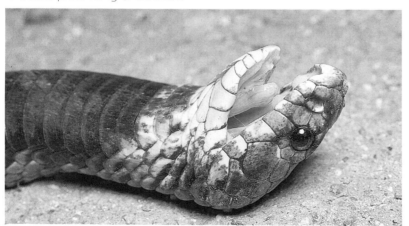

Whose eyes are these?

Another way to look larger than life is to have a big pair of eyes. Some moths will spread their wings if disturbed. The underwings have two big eye spots which look like part of a large face. The puss-moth caterpillar inflates its body until two large false eye spots appear over its head, giving it a fierce face. One South American frog has large eye spots on its backside. When threatened, it raises its rump to its attacker.

Did you know?

Ladybirds produce poisonous blood from their knee joints when threatened.

Some fish give a predator a short sharp shock — an electric shock.

Some plants can warn other plants of approaching predators by means of chemicals which they waft through the air.

More about animals in armor (pp.22, 23); warning colors (pp.56, 57); camouflage (pp.52, 53)

Mimicry

monarch

viceroy

Look-alikes

Some animals, called mimics, look so like other animals or plants, their models, that the two creatures are easily confused. These two butterflies are from different families. The one on the left is the monarch butterfly. Caterpillars of the monarch butterfly feed on poisonous milkweed plants, and, as butterflies, their flesh is poisonous to their predators. It makes the predators very ill, but does not kill them. Once a predator has tasted a monarch, it wants to avoid eating another one. It can easily recognize other monarchs by their bright colors. The butterfly on the right is the viceroy. It is harmless, but predators avoid it, mistaking it for the monarch.

Acting the part

An animal may mimic its model in behavior to improve the deception. Perhaps the most surprising wasp mimic is the wasp beetle. It does not have transparent wings, but to look more convincing, it moves in short darting runs like a wasp.

Ant-mimic spiders have slender antlike bodies. They hold their front legs up like feelers and dash to and fro like the ants they are hunting.

Stick caterpillars are colored like twigs. To improve the deception they grip the branch and hold their stiff bodies away from the branch to look like twigs. They may not move all day.

Warning stripes

The bright black-and-yellow stripes of the wasp warn of its vicious sting. Lots of different insects have the wasp's colors — hoverflies, lacewings, beetles and even a moth which has a striped body and transparent wings.

◄ Wasp-mimic moth.　　　　▼ Wasp beetle.

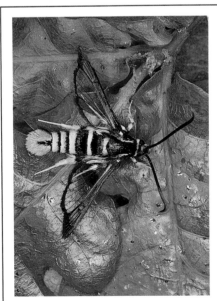

▲ Red and yella', kill a fella'.　　▼ Red and black, friend of Jack.

▲ Some plants look like animals. Bee orchids have furry bee-colored petals, and give off a scent like that of a female bee. The orchid needs to receive pollen from another orchid flower in order to produce seeds. Male bees are attracted to the flower and try to mate with it. In the process, the orchid pollen sticks to their bodies. Pollen from other orchids the bees have visited rubs off on the orchid.

Red and black, friend of Jack

In North and South America there are many different species of coral snake. Many of them are very poisonous. Coral snakes have bright red, black and yellowish white warning stripes. Harmless kingsnakes living in the same area also have red, black and whitish stripes. Local people distinguish between the harmful and harmless snakes by the order of colors of the stripes.

Dangerous deceits

Sometimes only one part of an animal mimics something else. The alligator snapping turtle has a fleshy red flap of tissue on the floor of its mouth, which it wiggles so it looks like a worm. The turtle opens its mouth and wiggles its lure. Small fish, thinking they can catch the worm, swim right into the turtle's open jaws.

Anglerfish use the same deceit. Wiggly lures grow out from their chins or dangle above their mouths. Many have huge mouths. As soon as a small fish comes within reach, attracted by the lure, the mouth opens wide and the prey is sucked in.

— *Did you know?* —

The lantern bug of South America looks just like a miniature caiman (a South American alligator). Is it mimicking the caiman? Answer on page 152.

Some caterpillars coil round to look like snails when they are resting.

Tiny *Extatosoma* mantids look like ants when young. As they get too big for this disguise, their bodies change color and shape and they mimic scorpions instead.

More about warning colors (pp.54, 55); anglerfish (p.119) look-alikes (p.25)

57

Tool users in the animal world

Do animals really use tools?

Many people have seen or heard a song thrush banging a snail shell against a large stone. Song thrushes often use the same stone over and over again. You can easily recognize the song thrush's "anvil" by the many pieces of broken shell scattered around it. Song thrushes behave like this because they cannot get the snails out of their shells in any other way. Is the song thrush using the stone "anvil" as a tool? Do animals, other than humans, really use tools? Some scientists think they do. A number of wild animals have been seen using simple objects to help them do certain jobs.

▲ Woodpecker finch using a twig to dig an insect out of the bark.

Making beaks and arms longer

The woodpecker finch lives on the Galápagos Islands in the Pacific Ocean. It feeds on small animals including insects and spiders. It looks for its prey on the bark of trees. However, insects and other small animals can easily hide in the cracks in the bark. When they do this, they are difficult to reach. The woodpecker finch solves the problem by poking a twig into the crack to "winkle" out its food. While it gobbles up its catch, it drops its twiggy tool. When it is ready to search for more food, it holds the tool in its beak and starts poking between the cracks again.

Termite "fishers"

Wild chimpanzees have been seen catching termites with tools they have made. They use a variety of objects, including twigs and strong grass stems. The chimpanzees modify the tools by chewing the ends to make a kind of brush. They then push the tool into a termite mound in search of insects. Any termites disturbed by the chimpanzees bite the brushlike end of the tool and hold on. All the chimpanzee has to do is pull the tool out and eat the termites on the end. Using this tool, chimpanzees can catch food otherwise out of reach.

▲ The Egyptian vulture feeds on different kinds of food, including eggs. It can easily break open small eggs by picking them up in its beak and then dropping them on the ground. It deals with bigger eggs in a different way. It doesn't take the eggs to a stone. Instead, it drops a stone on the egg. It usually takes several attempts to break the shell, but the vulture nearly always succeeds. If a stone of the right size isn't available, birds may go some distance to find something they can use.

The bearded vulture drops bones from a great height. The bones shatter on the rocks below. The vulture can then feed on the marrow inside the broken bones. Sometimes the vultures drop tortoises to break their hard shells.

A sticky problem

A few animals are known to use sticky materials for gluing other things together. The satin bowerbird lives in north-eastern Australia. It squeezes bark from a particular tree to extract a sticky fluid. It mixes this with its own saliva. The mixture is then used to plaster the walls of its bower home. This bird also paints the inside of its bower with juice from wild berries. It uses a piece of bark as a paint brush.

Marine strongman

The sea otter lives in the Pacific, off the west coast of North America. When it gets hungry it dives to the sea bottom. It returns to the surface carrying a rock in one paw and a shellfish in the other. The otter then floats on its back and breaks the shell-fish open by banging it against the rock held on its chest. Sometimes this hammering can be heard far away.

Did you know?

Harvester ants, in southeast Asia, use bits of stick as plates to carry home food, or to soak up mud and sand. By doing this they can carry 10 times more food than if they swallowed it.

In southeast Asia some tailor ants squeeze their larvae and extract a glue. They use this to stick leaf edges together.

The dwarf mongoose, in Africa, holds an egg between its back legs, and breaks it by flicking it backward against a rock.

More about animal skills (pp.60, 61); Galápagos finches (p.21)

59

Animal builders

Buildings large and small

Animals build for different reasons. They build homes in which to live and nests in which to rear their young. Some animals even build traps to catch other animals. The types of structures animals build vary in size and complexity, and they are made from many kinds of building material.

Some animals build tubes and tunnels where they can live. Certain marine mollusks can bore holes in hard rock. Worms living on the seabed often make complicated and delicate tubes which protect their soft bodies from predators. Other animals burrow in the soil. Moles, rabbits and badgers are good excavators. They are able to dig long, complicated underground passages. There are some birds which are also tunnel experts.

Insects are skilled architects and builders. Bees make intricate honeycombs based on sound mathematical principles. Weaver ants from tropical southern Asia glue leaves together by using their own larvae as a weaver uses a shuttle. Termites are even more skilled as planners and builders. They make huge mounds that have complicated air-conditioning systems.

▲ Termite mounds are often very big structures made of sand or clay. The one in the picture is built by an African termite called Macrotermes. Inside there is a complicated system of ventilation shafts which help air circulate through the mound. This species of termite also grows fungus gardens. The humidity in the gardens must be carefully controlled. Some termites living in tropical rain forests put roofs with overhanging eaves on their buildings which make them look like pagodas. The roofs keep the heavy rain out. Other rain forest species build the whole mound in the shape of a mushroom.

Honeybees: the mathematical builders

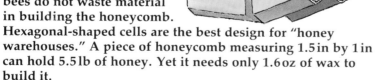

Honeybees build the cells of their honeycomb very accurately. Each cell is hexagonal in cross-section. The hexagonal units fit together neatly, so that there are no spaces left between them. This means that the bees do not waste material in building the honeycomb. Hexagonal-shaped cells are the best design for "honey warehouses." A piece of honeycomb measuring 1.5 in by 1 in can hold 5.5 lb of honey. Yet it needs only 1.6 oz of wax to build it.

Honeybees are able to measure precisely. Each hexagonal cell in the honeycomb slopes downward from the opening to the base. The slope is always 13°. Can you guess why? Answer on page 152. The distance across a worker cell is 0.2 in. It is 0.25 in for a drone cell. Each cell wall is 0.003 in thick.

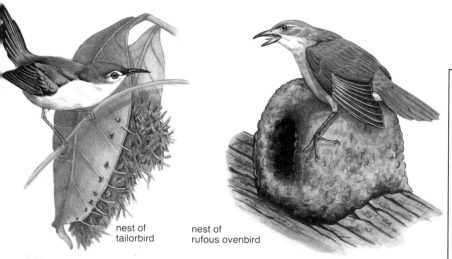

nest of
tailorbird

nest of
rufous ovenbird

From potter to weaver

Birds are some of the most ingenious builders in the animal
world. Some dig holes in the ground. Others bore holes in
trees. They use all kinds of materials and building
techniques. There are basket makers, weavers, carpenters
and potters. The nests they build vary in size and
complexity. The vervain hummingbird from the West
Indies builds a nest no bigger than half a walnut shell. The
brush turkey from Australia, on the other hand, makes a
mound several yards in diameter in which the eggs are
incubated. The social weaver bird from Africa makes a
communal nest in which fifty or more pairs of birds live
together.

Male bowerbirds construct very elaborate structures with
which to attract the females. Some bowers are built in the
shape of a circle with a pole in the middle. Others are made
to look like an avenue of small pillars. The orange-crested
bowerbird from New Guinea makes a stage-shaped bower
and decorates it with colored berries and beetles.

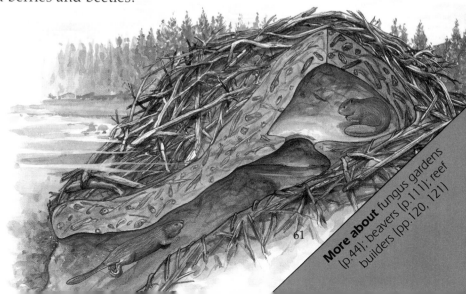

► Beavers are skilled engineers.
They cut down trees and use the
timber to build dams and lodges.
They construct canals along
which they float the trees to the
building site. They build a dam
across a stream which creates a
pond. The beavers then build a
lodge in the middle of the pond
where they can live in safety.

More about fungus gardens
(p.44); beavers (p.111); reef
builders (pp.120, 121)

61

Territory

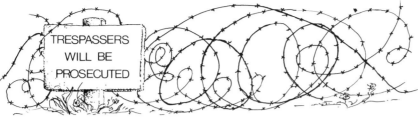

You have probably seen this sign before. We don't like other people walking on our property, so we put up signs and build fences and walls to keep them out. Many animals also like their own property. They don't usually build walls around it, but they do mark its boundaries by using all kinds of signs. We can't easily see them, but they are there for other animals to look out for, and they give the same message: "Keep out."

The home range

Many animals spend the whole of their lives in one particular area and they don't usually move away from it. They find their food there and they sleep there. They may even have "lookout" posts in the area, and places where they can hide when danger threatens. An area like this is called the home range. The size of the home range varies from one species to another. For some species it is very big, and for others it may be quite small. Its boundaries are not usually defended, and other animals are allowed to enter.

Territory

Within the home range, animals may claim a special area which they defend against other animal trespassers, especially members of their own species. This area is called a territory. It can be defended by a single animal, by a pair or even by a group of animals. It may be defended all year round, or only during the breeding season.

Territory "flags"

The African male agama lizard has a bright-red head and an orange tail which act as his advertising flags. The bright colors and his fierce behavior tell other males to "keep out." Many male animals advertise themselves by their bright colors or by behaving in a certain way. This is how they keep their territory. Male robins claim their territory by singing loudly. This drives other males away. However, the male robin sings only on his own property.

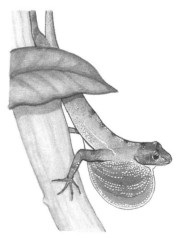

▲ Anolis lizard displaying throat pouch.

◄ Territories are usually defended by males. A bull elephant seal will attack any other male which enters his territory. He may even kill the trespasser to defend his harem of females.

Mock fights

Quite often male animals claim and then defend their territory without hurting each other. They take part in "mock fights" and the winner holds on to the territory. Antelopes and deer do this. At the beginning of the breeding season, the males fight among themselves. They fight in pairs by locking horns or antlers together. They push each other around until one gives up and is chased away. The strongest male wins and claims his territory. He also wins the collection of females that goes with it. Other males sometimes challenge him, and he may have to defend his territory and his harem of females.

Thinking through your nose

Although we can't see them, territories do have definite boundaries and animals mark these in different ways. Hippopotamuses spray their territory with liquid dung. It works like an aerosol spray. By waving their tails in the spray they mark the surrounding vegetation. The message they leave behind warns other hippopotamuses to "keep out." Tigers urinate on trees and other plants. The marked trunks act like fence posts telling other tigers to stay away. Bush babies and tarsiers even urinate on their hands and then spread the "message" by climbing through the branches.

Hear all about it

Animals living in thick forest find it difficult to advertise themselves and their territory. This is because they are not easily seen. They solve the problem by shouting. The sound of their screams can be heard over a mile away and it warns other groups to keep their distance. Gibbons and howler monkeys do this. So do many birds, including parrots and turacos.

War paint

A famous scientist called Konrad Lorenz was puzzled by the very bright colors of many coral reef fish. He did some experiments to find out why they were so brightly colored. He discovered that the special colors of each species act like a flag, telling other fish of the same species to keep away. The "war paint" worn by each fish helps it to keep its territory and stops trespassers from entering.

63

More about communication (pp. 72, 73); hearing, seeing, smelling (pp. 50, 51); animal defenses (pp. 54, 55)

Display and courtship

Courtship

Most animals need to find a mate before they can reproduce. Courtship behavior is used by male animals to attract a female and to get her into the mood to mate. Often several males compete for the attentions of a female. The male who puts on the most impressive display wins the lady.

Dressing up

Many animals use color to attract a mate. Many male birds have a brighter plumage in the spring. The breast of the male stickleback turns red in the spring. The bright color attracts females and also serves as a warning to other males not to come too close. Many courtship displays have this double aim. Male sticklebacks will even attack red circles of cardboard in the mating season.

▲ Male birds of paradise grow special, long tail feathers which are used in a spectacular display which involves acrobatics, fluffing up of feathers and strange calls.

Serenades

Many animals use sound to attract mates. Male frogs "sing": many of them have large air sacs which make the sound louder. The females are attracted to the male frog with the loudest croak. A chorus of frogs can attract females from more than a mile away. Grasshoppers and crickets chirp a serenade. Mayflies and mosquitoes gather in huge clouds of dancing males. Their whirring wings attract the females, who are recognized by the different sounds of their wing beats.

Sight and sound

Often animals put on a special display to show off their bright colors. The peacock spreads its magnificent tail into a great fan. The frigate bird inflates its large, red throat sac. The throat sac also helps it to make a booming sound, which also helps to attract the attention of females flying overhead.

Dancing partners

Sometimes the courtship display is in the form of a dance. The male stickleback performs a zigzag dance in front of the female.

The scorpion grabs his female by the pincers and dances her across the ground. Before the dance, he wraps his sperms in a tiny packet which he leaves on the ground. During the dance, he guides the female over the packet so she can pick it up.

Waving the flag

Some male spiders wave their legs in a form of semaphore to signal to the female that they are mates and not meals. The male fiddler crab has one claw which is much larger than the others. He waves this claw like a big white flag to attract the females to his burrow.

Giving presents

Many animals give their intended mates presents of food. Terns offer fish, while the American roadrunner gives his mate a lizard, which she continues to hold in her beak during mating. Roadrunners are awkward birds, and some people think the female uses the big lizard to help her balance while mating!

Some spiders give their mates food, wrapped up in silk. The female spider is much bigger than the male, and would normally attack and eat him if he came too close. The food parcel acts as a distraction while he mates with her. Spiders who cheat by giving their mates an empty parcel often end up as dinner.

The lure of perfume

All animals give off special scented chemicals called pheromones. Many male moths have huge antennae loaded with smell sensors. They can detect the scent of a female from a couple of miles away. Some animals can tell whether a female is ready to mate by her smell. You often see courting mammals sniffing near the tail of their partners.

▲ The male bighorn sheep sniffs near the tail of the female to discover whether she is ready to mate.

▲ Like these great crested grebes, many animals have courtship displays involving both partners.

—*Did you know?*—

During courtship, snakes tickle each other by rubbing their scales up the wrong way.

Mating snails inject chalky love darts into each other while courting.

The female baboon gets a very swollen bare pink bottom when she is ready to mate. She stands with her back to the male and her tail raised to show off her bottom.

More about how animals smell (p.51); how animals communicate (pp.72, 73); animal reproduction (pp.66, 67)

65

Reproduction

Why multiply?

No animal or plant lives for ever. Sooner or later it dies; it is eaten, has an accident or simply ages — its body wears out and becomes less efficient as it grows old. All plants and animals reproduce. They produce more plants or animals like themselves.

Simple multiplication

Very simple creatures reproduce by dividing in two. A few bigger, more complicated animals can also divide in two. Some sea anemones pinch in their bodies down the middle until they become two separate animals. The green hydras which live in ponds produce buds which grow rings of tentacles. The buds later drop off as baby hydras.

Advanced multiplication

For bigger and more complicated animals or plants it is difficult just to divide in two. Imagine trying to divide a human being in two. How can a tree divide in two when the two halves cannot move apart?

Most animals and plants reproduce sexually. When they reproduce, they make special sex cells. In animals, the male sex cells are called sperms, and the female sex cells are called eggs. Each sex cell contains half the information needed to make a new animal. During "fertilization" a sperm meets an egg and joins with it to form one new cell. This cell contains all the information for making a new animal. The cell then divides again and again until a new animal is formed.

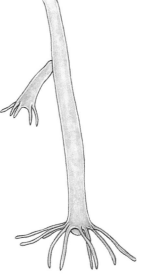

▲ Hydra budding.

Taking chances

Many animals mate in water. The sperms have tails and can swim to the eggs. The eggs attract the sperms by giving off special chemicals. Frogs and fish mate in water. This process is very wasteful: lots of eggs and sperms never meet each other, but drift away on water currents or get eaten by fish. Very large numbers of eggs and sperms have to be produced to make sure a new generation is made.

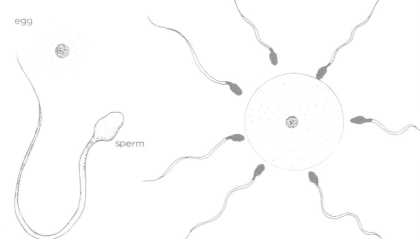

egg

sperm

◀ In order to reproduce, sperms have to meet the eggs. This happens during mating, when a male animal and a female animal come together and shed their sex cells at the same time.

Playing safe

Sperms need water to swim in. Some animals which live on land, like frogs and toads, return to the water to breed. Other land animals use internal fertilization. The male has a special tubelike organ, called a penis, which he uses to inject sperms into a special opening in the female. The sperms fertilize the eggs inside her body. Mammals and reptiles have penes, and most male insects have some sort of organ for transferring sperms to the female. Birds do not. With internal fertilization there is a much better chance of a sperm meeting an egg, so far fewer sex cells are produced.

Eggs inside and outside

▲ Rattlesnake giving birth.　　▲ Snakes hatching from eggs.

Some animals which reproduce by internal fertilization keep the eggs inside the mother's body in a special hollow, called a womb, until they have developed into baby animals. Then they are pushed out — they are "born." This happens in mammals, rattlesnakes and some lizards. In other animals, like birds, insects and most reptiles other than rattlesnakes, the eggs develop a tough waterproof shell. They are then released to the world outside. The baby animal goes on developing inside the egg until it is ready to hatch.

▼ Female trout laying eggs for male to fertilize.

Life before birth

While the baby animal is developing inside the egg or inside its mother's womb, it is called an embryo. Most eggs contain a bag of food called the yolk, which the embryo slowly absorbs. Baby mammals have an even better food supply. They are attached to their mother's womb by the umbilical cord. In this cord the embryo's blood vessels are very close to those of the mother, and they get food and oxygen from the mother's blood.

—Did you know?—

The ocean sunfish may produce 300 million eggs.

A newly hatched sunfish is 1,575 times smaller than its mother.

An ostrich egg is so strong that a human adult can stand on it without breaking it. Why does it need to be so strong? Answer on page 152.

More about how animals find mates (pp.64, 65); how baby animals grow up (pp.68, 69); animal parents (pp.70, 71)

Metamorphosis

Metamorphosis

A human baby looks very much like a tiny adult, but many baby animals do not look anything like their parents. They undergo a dramatic change during the development from egg to adult. We call this change in form "metamorphosis." The young animals below are very different from their parents. Can you guess whose babies they are? You will find the answers on page 152.

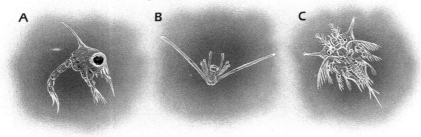

A   B   C

A change of diet

The young of some animals are very different from their parents and are called larvae. Butterflies and moths are elegant flying insects, but their larvae are slow, fat caterpillars. The adults and young feed on different things. Caterpillars feed on plants. They have hard, horny jaws for cutting up the leaves. The adults feed on nectar, which they sip through a long coiled tube called a proboscis.

The caterpillars are perfectly designed for growing fast. They have simple bodies, that are just fat food processors. As they grow, they moult often. They can move just enough to find another leaf to eat. The adults can fly far and wide in search of mates. They may lay their eggs some distance away. This helps the species spread to new places.

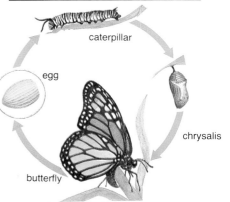

caterpillar

egg

chrysalis

butterfly

Reshuffling the building blocks

How does a caterpillar change into a butterfly? When it is fully grown, the caterpillar forms a hard case around itself. It is now called a pupa or chrysalis. Sometimes it spins a silken cocoon around the case. Inside the case, the caterpillar's tissues break down and the "building blocks" of living material are rearranged to form the adult butterfly or moth. After a time, the pupal case cracks open, and the adult emerges. At first the wings are crumpled where they have been folded inside the pupa. Soon they stretch out as blood is pumped along their veins. Then they harden in the air.

Growing problems

Insects and crustaceans grow in a different way from other animals. They are covered in a hard shell or skin, called the cuticle, which limits how much they can grow underneath. Every so often, the shell splits open and comes off. Below is a new soft shell, which hardens quickly in air. Before it does so, the animal quickly puffs itself up with air, so the new shell sets as a bigger one. Then the insect or crustacean has room to grow under the shell.

Changing slowly

Not all insects go through dramatic changes like the butterflies and their caterpillars. Some newly hatched insects, like grasshoppers and locusts, do not have proper wings. They are called nymphs. Every time they molt, the new cuticle forms around the growing wing buds. The tiny wings get bigger every time, until, at last, the insects are ready to fly.

Living like their ancestors

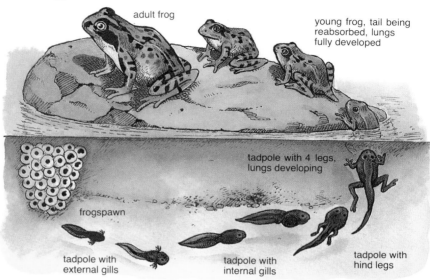

adult frog

young frog, tail being reabsorbed, lungs fully developed

tadpole with 4 legs, lungs developing

frogspawn

tadpole with external gills

tadpole with internal gills

tadpole with hind legs

The vertebrates evolved in the sea. First came the fish. From them, the amphibians evolved. The amphibians have legs and can live on land. Frogs and toads are amphibians. They catch insects with their sticky tongues and have lungs for breathing air. But their larvae, or tadpoles, are still fishlike, as their ancestors were, with long, flat tails and no legs. They breathe with gills, like the fish.

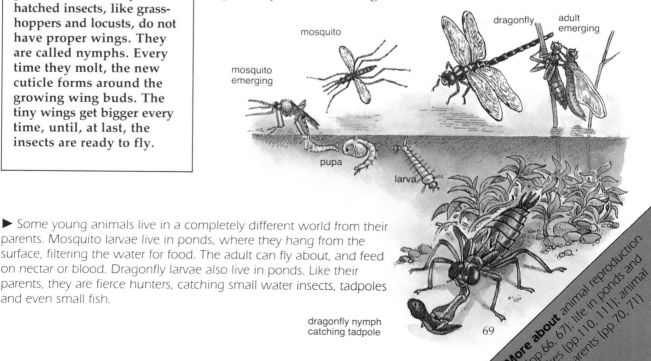

mosquito

dragonfly

adult emerging

mosquito emerging

pupa

larva

dragonfly nymph catching tadpole

▶ Some young animals live in a completely different world from their parents. Mosquito larvae live in ponds, where they hang from the surface, filtering the water for food. The adult can fly about, and feed on nectar or blood. Dragonfly larvae also live in ponds. Like their parents, they are fierce hunters, catching small water insects, tadpoles and even small fish.

More about animal reproduction (pp.66, 67); life in ponds and lakes (pp.110, 111); animal parents (pp.70, 71)

69

Parental care

The numbers game

Have you ever wondered why some animals, like the cod, produce millions of young, while others, like the elephant, produce only one offspring at a time? Once the cod's eggs are fertilized, neither parent has anything more to do with them. The young cod are very small, and are eaten by many different predators. Because of this very few survive. However, the baby elephant stays with its mother for several years. A much higher proportion of baby elephants survive to become parents themselves compared with the cod, where perhaps only one young survives from several million eggs.

◀ New-born harvest mice in nest.

Helpless babies

Where vertebrates produce large numbers of offspring, their young are often small and helpless at birth. If they were to be born any bigger, the mother would not have enough room inside her to produce so many. Imagine an elephant trying to produce several baby elephants at once. Many baby birds and mammals are naked at birth — their feathers or fur grow later. Often their eyes and ears are still closed and they are quite helpless. They do not have warm coats, so they need to be close to their mother's warm body. Most mammals and birds make nursery nests, lined with soft grass or feathers.

Learning who's who

Many animals learn to recognize their own species from learning the characteristics of their parents. Goslings brought up in a swan's nest think that they are swans. When they find a mate, they will court swans instead of geese.

Ready to run

The baby giraffe is born with a full coat of fur and is able to walk within a few hours of its birth. It has to be able to keep up with the herd right from the start. The young of ground-nesting birds like chickens and lapwings hatch fully feathered. They can run around and find their own food within hours of hatching.

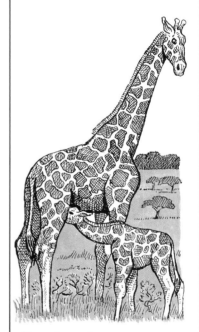

Food on tap

All baby mammals stay with their mother for a while after their birth because, at first, they feed on her milk. Once they can feed themselves and no longer need the milk, they are said to be weaned.

Parents on the move

Not all animals stay in or near a nursery nest. Some spiders and scorpions carry their young on their backs, and so do some frogs. Crabs and lobsters carry their eggs and young in a fold of the abdomen, or among the bristly hairs on their back legs.

▲ Scorpion with babies.

Playing for the future

Even after they are weaned, some baby mammals stay with their mothers for a time. Carnivorous mammals like foxes and cats learn how to hunt. They do this by watching their parents. Lions and cheetahs bring back their prey alive so that the young can learn to kill. Foxes hold food just out of reach of their babies, then jerk it away as if it were still alive. Play is useful. It helps young animals become good at judging distance for pouncing. It also teaches them to react quickly. Baby animals at play seem to be rather rough.

▶ Baby bees and ants are cared for by the whole community. Bee larvae are kept in special cells in the comb, and are fed and cleaned by worker bees.

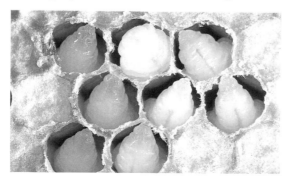

Colorful clues

Among birds, you can tell which parent cares for the young by the color of its coat. If a bird has to spend a long time sitting on the nest, it must be well camouflaged. Compare the gaudy plumage of the male pheasant with that of his dull mate. Who incubates the eggs? In most thrushes both the male and the female are dull-colored. What does this tell you? Answer on page 152.

Caring creepy crawlies

Parental care is found in most animal groups almost all over the animal kingdom. Centipedes lick their eggs free of fungus spores and curl around the eggs to guard them. Parent bugs stand guard over their eggs and young. Spiders spin special webs, called cocoons, to protect their young.

— Did you know? —

A new-born kangaroo, or joey, is only 1/30000th of its mother's weight.

A new-born panda grows so fast that its weight increases twenty-fold in the first 8 weeks of its life. If a human baby grew at the same rate, it would weigh 140lb when it was just 8 weeks old.

More about why animals produce so many young (p.20); animal reproduction (pp.66, 67)

Communication

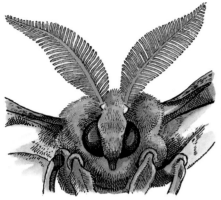

◀ This male moth has large, feathery antennae. They look a bit like tiny television aerials. They pick up chemical signals from a female moth. Often they can pick up the signals over very long distances. Chemical signals are called pheromones. Crabs, barnacles, spiders and insects all use them for communication. Mammals also use them for attracting a mate.

Animal communication

There are thousands of different human languages. They are all very complicated. We have to learn how to use a language correctly. This usually takes a long time. Animals do not have languages like ours, but they can communicate. For example, they use visual signals, sounds, smell, touch and even chemicals to tell each other what is going on. You might be surprised to learn that whales can sing and bees can dance.

Follow that tail

Animals which live together in groups or herds often flash signals to each other. White bottoms or tails with white underneath are good for giving a quick "message." A sudden white flash is a warning of danger. The signal is given when animals turn and start to run. Antelopes and rabbits signal like this. The African warthog runs away from danger with its tail held straight up. The tail acts like a flagpole and points the way for others to follow. The ring-tailed lemur from Madagascar uses its tail like a flag. The lemurs in a group feeding on the forest floor keep their tails upright so they can all see each other. This keeps the group close together.

A

B

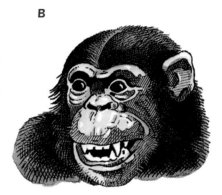

C

▲ Chimpanzees make different faces to show their feelings. Here are three pictures showing a chimpanzee in three different moods. Can you match A, B and C against the feelings of excitement, fear and happiness. Answers on page 152.

Open wide!

Young birds communicate with their parents when they want food. They keep their mouths open wide so that their parents can pop the food inside. Often, the inside of the chick's mouth is brightly colored to make it easier for the parents to see.

Elephant "talk"

Scientists have now discovered that elephants make low-frequency sounds, too low for humans to hear. They may be used in communication.

Sea sounds

Whales and dolphins make all kinds of sounds. Some of these are used for navigation but others are for communication. Humpback whales sing long and complicated songs. Each whale has its own song which is different from those of other whales. They can be heard a hundred miles or more away. Whales keep "in touch" by singing to each other. It may be that one whale can talk to another whale on the other side of the world!

Listen to me!

Shouting is a good way to communicate, especially if you live in thick vegetation. Groups of monkeys and flocks of birds shout to each other as they move among the branches. This keeps them together and stops anyone getting lost. Each species has its own "voice" which members of its group can recognize. Insects use sound to communicate. So do frogs and toads.

◀ Kiss me quick! Prairie dogs are small rodents from North America. They live in large groups in underground burrows. When two prairie dogs first meet, they kiss each other to find out if they are from the same group. If they recognize each other they start grooming each other's fur. If they are strangers they fight and the trespasser is sent away to find his own group.

Honeybees have a complicated dance language. They can tell each other where to find a new supply of food, how far away it is from the hive, and which direction to fly to find it. Bees have different languages. Russian bees cannot understand Italian bees and vice-versa.

Did you know?

Beavers slap the surface of the water with their tails as an alarm signal.

A rhinoceros has scent glands on its feet which it uses to pass on chemical messages to other rhinoceroses.

Humpback whale songs have made the record charts in the United States.

More about hearing, seeing and smelling (pp.50, 51); attracting a mate (pp.64, 65); whale songs (p.81)

Rhythms and "clocks"

Various rhythms

Most animals and plants show some rhythmic behavior. They do certain things at specific times and they repeat them regularly. Some rhythms follow a twenty-four hour pattern. Others occur once every month. Birds and other animals migrate from one place to another every year. Such journeys are examples of annual rhythms.

Day and night rhythms

Most animals become active at particular times every twenty-four hours. Some move about and feed during the daytime. These are called diurnal animals. Nocturnal animals are more active at nighttime. Most moths, owls and bats are nocturnal. Within any twenty-four hour period there are often peaks of activity. For example, troops of gibbons in the rain forest canopy are very noisy at dawn. So are many birds. We even talk about the bird "dawn chorus." Gorillas and chimpanzees feed in the morning and rest in the heat of the afternoon. Howler monkeys shout at dawn and in the evening.

Many plants hold their flowers and leaves in one position during the day and in another position at night. The blooms of the moonflower plant open at night and close during the day. They are pollinated by moths and bats. The leaves of many plants change their position throughout the day as they follow the Sun's path across the sky.

Regular deep sleep

South American humming-birds are very small and very active. They use up enormous amounts of energy. They also lose a lot of heat across their body surface. This creates many problems. They cannot supply their bodies with enough energy to keep active for more than twelve hours at a time. Because of this, they go into a deep sleep, rather like hibernation, for twelve hours every night. In this way they keep their energy demands under control.

South Pole roundabout

Karl Hamner wanted to learn more about daily animal rhythms so he went to the South Pole and did an interesting experiment. He put some small animals on a revolving turntable and watched their behavior. The turntable went around at the same speed as the Earth, but in the opposite direction. The animals didn't experience night or day, or even time. They were "standing still" in space. Even so, they still showed all their normal twenty-four hour rhythms. What do you think Karl Hamner concluded? Answer on page 152.

stoma

▲ Green leaves have tiny pores on their surface called stomata. The stomata open and close following a daily rhythm.

► At the end of May every year, herds of wildebeeste begin their migration across the East African plains. They must start the journey on time to reach their feeding grounds. This yearly rhythm needs careful timing.

Moon rhythms and the tides

Other animals and plants show behavior patterns which are related to the phases of the Moon and the rise and fall of the tides. Some fiddler crabs change color between high and low tide. They also show different activity patterns. Crabs come out of their burrows to feed at low tide. When the tide returns, they go back to their holes to hide. The palolo worm from the South Pacific is even more remarkable. Every year at dawn on one particular day in October the surface of the sea becomes frothy with writhing worms. The palolo worm is spawning and its cycle is linked to the Moon's phases. The day before there is no activity and the day after it is all over. Another year passes before the palolo worm spawns again. The life cycles of many seaweeds are also controlled by the rhythms of the Moon and tides.

Migratory and seasonal rhythms

Many animals and plants show behavior rhythms which are linked to changes in length of daylight, temperature and even food supply. Butterflies, birds, mammals and even some reptiles make yearly journeys called migrations. These always take place at the same time each year. The animals which take part in these migrations seem to have a good sense of time. For example, between June and August every year, most of the world's female leatherback turtles gather at the Trengaunu Beaches in Malaysia to lay their eggs. How do they time this event? Answer on page 152.

Many trees lose their leaves in winter. This is a rhythmic activity related to the fall in temperature and light. In tropical countries other trees behave in a similar way, but this rhythm is linked to hotter temperatures and the dry season. Some animals start to hibernate as winter approaches. This shows another annual rhythm.

Timekeepers

Some mangrove swamp snails are remarkable timekeepers. At low tide they climb down from the trees to browse on the exposed mud. However, they need to return to the safety of the trees before the tide comes in. They are slow movers, so they have to start their journey back before the tide begins to turn. If they start back too late, they will be caught by the rising water. Their internal "clock" allows the snails to forecast the tide's return. Why do they reset their "clocks" every day? Answer on page 152.

Animal "clocks"

Not all animal rhythms are controlled by changes in their surroundings, such as day or night, tidal movements or the seasons. Many rhythms still occur even when animals are put in surroundings which do not change. Karl Hamner discovered this at the South Pole. Animals which do this must have some way of telling the time. Scientists now think these animals have some kind of "clock" inside their bodies. They are not sure how such "clocks" work, but chemicals called hormones may be important.

More about migration (pp.78, 79); mangrove swamps (pp.114, 115); hibernation (pp.76, 77, 96)

The big sleep

▶ In winter, water freezes in the soil, making it difficult for many plants to survive. The problem of survival is solved in different ways. Some trees and shrubs lose their leaves and grow new ones in the spring. Some plants "die back" to the level of the soil. Their shoots grow again in warmer weather. Many plants pass the winter as seeds, bulbs or corms. These are all dormant stages which need no energy. They are the inactive part of the life cycle of plants.

Super seeds

Harold Schmidt found some Arctic lupin seeds in the frozen soil in the Yukon, Canada, in July 1954. The seeds were tested scientifically and found to be at least 10,000 years old. The seeds were germinated in 1966, 10,000 years after they fell off the plant. They had lain dormant all this time and had survived the coldest winters.

▲ Fish, amphibians and reptiles are cold-blooded. Their body temperature is always the same as their surroundings. As the outside temperature drops, they become sluggish and sleepy. The cold works like an anesthetic which slowly puts them to sleep.

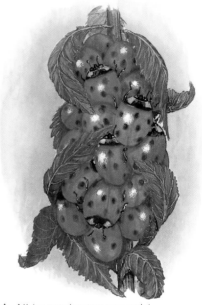

The problem of winter

Animals and plants which live in cold parts of the world face the problem of survival in winter. Food and water are scarce at this time of year, and keeping warm becomes difficult. Animals and plants solve the problem of winter in different ways. Many birds escape it altogether by migrating to warmer countries. They return in the spring. Some mammals, such as squirrels, store food in their larders during the summer and then visit these from time to time during the winter. Wolves grow a thicker coat to give them better protection against the cold winds. Shrews, voles and lemmings survive the Arctic winter by living and feeding beneath the snow. Some animals don't even try to find food in winter. Instead, they go to sleep. They hibernate.

▲ All invertebrates are cold-blooded. They survive winter in different ways. Insects sometimes pass the winter in the pupal stage and emerge as adults in the spring. Some butterflies hide themselves away in a sheltered spot and slowly go to sleep as the temperature drops. This sleeplike state is called torpor. Flies, bees and ladybirds do the same thing, often huddling together in large numbers.

Mammals

Mammals are warm-blooded. They have a high body temperature which is controlled by a kind of "thermostat" in the brain. Keeping a high body temperature needs plenty of energy, especially for small mammals. This means they must eat a lot of food. Winter is a problem because food is scarce, and so some mammals hibernate. In order to survive winter, hibernating mammals must use less energy, so they must be less active. They limit their activity by turning down their "thermostats," and lowering their heartbeats and breathing rates. Now the body "ticks over" very slowly using very little energy. The fat stored up in the summer provides all the energy needed. Some mammals even have a kind of "antifreeze" in the blood to stop it freezing up.

Some mammals go into a "temporary" hibernation from which they wake up from time to time. Bats do this. Bears do something different again. The European brown bear and the American black bear simply go into a deep sleep. It isn't real hibernation because the body temperature never drops below 60°F. Female bears even give birth to their young during this time.

▲ The dormouse has a "thermostat" in its brain which controls its body temperature, keeping it at just under 104°F. As winter approaches, the dormouse builds a nest, curls up in a ball, and goes to sleep. Can you think why it curls up? It turns its "thermostat" down so its body temperature drops to about 43°F. Its heartbeat slows from 300 beats to about 10 beats per minute. Now it can survive the winter by using very little energy. It gets what energy it needs from its stored body fat. In spring it turns its "thermostat" up again and starts to wake up. Perhaps the dormouse in Alice in Wonderland was hibernating!

The whippoorwill

Birds are warm-blooded. Not many birds hibernate, although some swifts and hummingbirds go into a deep sleep like hibernation to save energy. The whippoorwill, found in eastern North America, does hibernate. As winter approaches it builds up fat reserves to help it survive. Then, when the weather gets colder, it crawls into a sheltered hole and goes to sleep. Its body temperature drops to below 60°F. It stays like this for several months, slowly using up its food reserves. It needs about 0.35 ounces of fat for every 100 days of sleep.

The sleeping fish

The African lungfish survives drought in a remarkable way. If the lake in which it lives dries up, it burrows into the mud and surrounds itself with a waterproof cocoon. Breathing air through a ventilation shaft to the surface, it lives off its own muscle tissue. It soon revives when the rains come. Some tropical frogs and toads do the same thing. This kind of hibernation is called estivation.

More about hibernation (p. 96); life in cold climates (pp. 100, 103); coping with dry weather (pp. 96, 97)

77

Migration

On the move

Not all animals live in the same place all the time. Many animals live in one country in the summer and in another country, or even another continent, in the winter. Other animals spend the first part of their lives in one area, and their adult lives in another. These regular journeys are called migrations.

The Sun Seekers

The lands of the far north, especially the Arctic tundra, are favorite places for many animals to have their young. When the snow melts in the spring, there is plenty of new plant growth. In the melt-water pools millions of insects thrive, providing food for many birds. The cold polar waters are also full of fish and small crustaceans, which are food for many seabirds. Because the winters are too cold for most animals to survive, there are no large resident populations of animals here to compete for the food.

Many geese, ducks, waders and terns breed in the tundra, but spend the winter in warmer climates, often south of the Equator. Swallows and warblers, which feed on insects, migrate every year from tropical regions to breeding grounds in northern Europe and America.

In mountainous regions of the world, wild goats and sheep move up to high alpine pastures in the summer. In the autumn, they move back to the shelter of the wooded valleys where they feed in the shelter of the trees.

From land to water . . .

Some animals need special conditions for mating or egg laying. Frogs and toads are evolved from fish, and they still shed their eggs and sperms into water when mating. For most of the year they live on land, feeding on insects, but in the breeding season they gather together at favorite ponds to mate, often traveling a mile or more.

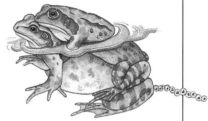

. . . and water to land

Other animals migrate from the sea to the land. The marine turtles evolved from land-dwelling ancestors. They return to sandy beaches to lay their eggs. Here the eggs incubate in the warm sand. When they hatch, the tiny turtles immediately make their way back to the sea.

◀ North American caribou spend the summer feeding on the tundra, but retreat hundreds of miles to the sheltered forests in the winter. Packs of wolves follow them.

Exploding populations

Sometimes animals multiply too fast, and run out of food in their home area. Vast numbers of animals then move out to find new feeding grounds. These one-way mass movements are called emigrations. They are common in lemmings, small vole-like animals that live in Scandinavia and Canada. In warmer climates, huge swarms of locusts also emigrate in search of food.

Following the rains

In the dry grasslands of Africa, and similar areas, large herds of grazing animals follow the rains, migrating in search of fresh, green vegetation. Elephants, antelopes and wildebeeste follow regular routes, hundreds of miles long, across the great plains.

Ocean wanderers

In the sea, whales migrate thousands of miles in search of food. Fish like herring and cod migrate to special shallow-water areas to breed. Salmon spend most of their adult lives at sea, but when the time comes for them to breed, they travel thousands of miles across the oceans to the rivers in which they were born, struggling upstream to lay their eggs. The young salmon return to the oceans to mature. Common eels make the opposite migration, spending their adult lives in fresh water, but returning to the ocean to breed.

▼ Insects also migrate. In Europe, large numbers of clouded yellows, red admirals and other butterflies frequently fly across the English Channel to the United Kingdom in the spring and summer. In the United States and Canada, monarch butterflies migrate some 1,250 miles south to Mexico and the southern United States in the autumn. In the winter, monarch butterflies rest, crowded together in favorite roosting trees. In the spring, they migrate north to breed.

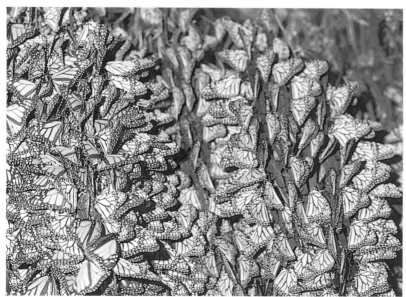

---*Did you know?*---

Most bird migration takes place out of sight of human eyes, 9,000 ft or more up in the atmosphere.

In Africa in the last century, a herd of migrating springbok may have contained one hundred million animals.

In the ocean, lanternfish migrate to the surface each night to feed, returning to depths of 1,367 yd at dawn.

More about migration (pp.75, 103, 104); how migrating animals find their way (pp. 80-83)

How animals find their way

Why do animals need to navigate?

Many animals are able to navigate accurately. They may have to find their way back home after hunting for food. They sometimes have to navigate to find a mate, or to escape from enemies. Some have to leave an area for a short time because the weather is very bad — for example, because the winter is too cold and harsh. Animals often navigate over long distances when they migrate. Migrations may involve journeys from one part of the world to another. Animals usually migrate when the seasons change.

The Sun is always up

It is important for an animal to know in which direction to travel. It is also important for the animal to know about the position of its body in space. Fish, insects and birds use the Sun to tell them about their own position. They also use gravity to tell them which way up they are. Other animals use chemicals to help them navigate. Salmon and eels migrate long distances. Chemicals in the Atlantic Ocean help them to find their way.

Local landmarks

During short journeys, animals often use local landmarks to help guide them. Insects, birds and mammals do this.

Electrical signals

Tropical lakes and rivers are often very muddy. Some species of fish which live in these conditions find their way about in a special way. The electric eel from South America sends out large numbers of electrical signals. It surrounds itself with these signals which make a special pattern. If the signals hit against something in its surroundings the pattern changes. This tells the eel that there is something close by. These signals can also be used to stun other fish on which the eel feeds.

◀ Bats navigate by sonar. They produce high-pitched sounds which bounce back off objects in their flight path. They have very large ears to catch these echoes. This system of navigation is called echolocation. Bats use echolocation to find their way even in complete darkness. It works like radar and is very accurate. Bats also use sonar to catch insects. There are bats from Central America that use sonar to find the fish on which they feed.

Singing whales

Whales and dolphins make sounds to help them find their way and to keep in contact. The sounds are usually high-pitched like the sonar of bats, although lower sounds are also produced. Dolphins seem to "talk" to each other by "whistling." Their high-pitched clicks help keep the group together. The humpback whale sings for long periods and it also sings different songs. A single song may last for thirty minutes. Whale songs can travel for hundreds of miles underwater. The songs may help in navigation during the whale's annual migration. They almost certainly help to keep the whales together. Only the male humpbacks sing, so their songs may help in finding a mate. Whales also follow the patterns of the Earth's magnetic field on the seabed. They follow these rather like cars on a motorway. Sometimes they come off the "motorway," crash on a beach and become stranded.

Insect navigation

Butterflies navigate by sight and smell. Species which migrate also use the Sun's position to help them find their way. The monarch butterfly travels more than 1,900 miles on its migration. It uses the Sun to help it. The Sun's position in the sky changes throughout the day. The monarch butterfly allows for this movement when it navigates. Not all migrating butterflies do this. The cabbage white and the red admiral butterflies also migrate over long distances. They do not seem able to compensate for the Sun's movement.

Dancing bees

Worker bees do a special dance on the face of the honeycomb when they want to tell each other about a group of flowers they have discovered. When it dances, the bee follows a path shaped like the figure 8. The line between the loops of the 8 acts as a pointer. It tells other bees how far to the left or right of the Sun they have to fly to find the flowers. The dancing bee also tells the others how far away the flowers are by wagging its abdomen. Quick wagging means the flowers are close to the hive. Slow wagging means they are a long way off.

— Did you know? —

The blue whale produces the loudest sound of any animal. It can be heard 525 mi away.

The oil bird from Venezuela flies in complete darkness in the caves in which it lives. It echolocates like a bat.

Some moths "jam" the sonar system of bats hunting them. This helps them escape capture.

Nose-leaf bats amplify the sound they produce, just like a megaphone.

More about bird navigation (pp.82, 83); animal senses (pp.50, 51); navigating in the dark (pp.126, 127)

Birds — master navigators

Long-distance travelers

Certain species of bird leave their winter feeding grounds and fly to summer breeding areas. This type of journey is called a migration. The birds may fly many thousands of miles during these journeys. After breeding they make the return trip back to the feeding grounds they came from.

▲ Arctic tern.　　　　　　▼ Migration route of Arctic tern.

Bird navigation

Arctic terns hold the record for the longest migration. The birds breed in the Arctic and the northern hemisphere. When breeding ends, they fly south to Antarctica where they spend the winter. Sometimes the birds fly for long distances over the open sea. They have no landmarks to guide them but they still manage to navigate accurately. After spending the winter near the South Pole, the birds fly north again to their breeding grounds. The round trip may be as much as 24,800 mi. It requires accurate navigation to fly there and back safely.

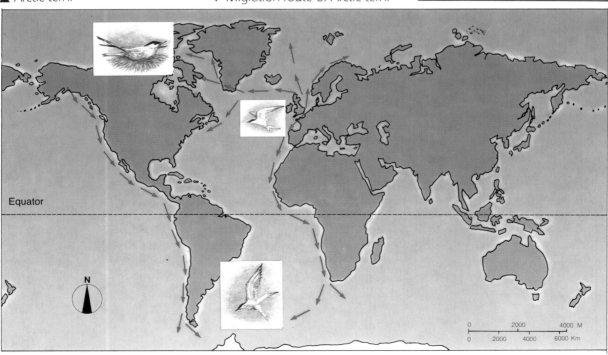

Equator

N

| 0 | | 2000 | | 4000 M |
| 0 | 2000 | 4000 | | 6000 Km |

How do birds find their way about?

Scientists have puzzled for a long time about how birds navigate. Some species follow clear landmarks which they recognize using their keen eyesight. For example, they identify rivers, estuaries and coastlines. This is called visual navigation. Birds have much more sensitive eyes than we do. They can see very fine detail on the ground below them, which may help them to find their way.

Navigating over long distances

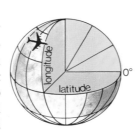

Birds flying over long distances probably navigate in a similar way to a trans-atlantic jumbo jet. In a jet, the captain uses a special kind of compass, a flight plan, and the computers on board to make sure the plan is followed accurately. The flight crew is able to tell exactly where the jumbo is at any time during the flight. The crew can tell the correct latitude and longitude. The captain also uses radio beams along the route.

A bird probably carries a "flight plan" in its brain. The brain also has a built-in clock, map and compass. All these things help birds to navigate very accurately. Birds use the position of the Sun and stars to help them find their way. The bird's brain allows for the Sun's movement when following its flight plan. This helps it work out its position of longitude. Some birds navigate during cloudy nights. It is likely that birds also use the Earth's magnetic field to help them navigate even when the Sun and stars are not visible.

The remarkable emperor penguin

During the breeding season, emperor penguins make several long journeys across the frozen Antarctic. The males and females separate at certain times, but they navigate so accurately that they are able to find one another at the end of each trip. There are few landmarks to help and the weather is poor. The birds may use the Earth's magnetic field to help them find their way.

— Did you know? —

Swallows, shearwaters and puffins return to the same nesting site every year.

Young wheatears make their first journey from Greenland to South Africa without the help of their parents.

Bar-headed geese fly over the Himalayas during their migration. They fly as high as 26,000 ft.

More about how animals navigate (pp.80, 81); migration (pp.78, 79)

Life in the water

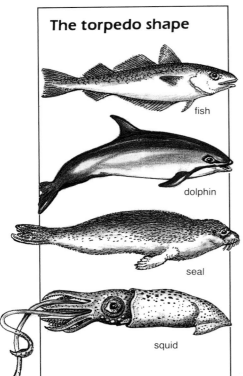

fish

dolphin

seal

squid

Many aquatic animals are torpedo-shaped. This makes them streamlined so they move easily through water. Whales and seals have extra layers of fat under their skin. The layers make their body outline smoother and more streamlined. Whales also have very short neck bones to make the front of the body more torpedo-shaped. Penguins become torpedo-shaped when chasing fish underwater. Octopuses and squids are not streamlined normally, but when swimming fast they change shape and become streamlined.

A watery world

Two-thirds of the Earth's surface is covered by water. There are many different kinds of animals and plants living in this aquatic habitat. Some animals and plants live on the surface of water. Others live below the surface at different depths. There are even animals in the deepest parts of the sea.

Floaters

Many animals and plants float on, or near, the surface of water. Plankton contains millions of tiny floating animals and plants. Animal plankton is called zooplankton. These animals often have unusual shapes which help them float. Some floating plants have air bladders which keep them afloat. Animals such as the Portuguese man-of-war have enlarged air bags which are used like sails. They are blown along by the wind.

Countershading

Next time you go to the fish shop, look more closely at the fish on sale. You will see that most of them are dark-colored on their backs, and lighter-colored underneath. This is called countershading. The pattern of dark and light-colors makes an aquatic animal difficult to see from above and below. Some fish swim upside down. In these fish the countershading is reversed. Can you think why? The larger animal in the picture, above, is called a dugong or sea cow. It lives in warm tropical waters. Does it follow the countershading rule?

◀ A hippopotamus may weigh more than 3 tons. Hippopotamuses live in tropical Africa. They are some of the biggest mammals. Even so, a fully grown hippopotamus is able to lie submerged in water without being seen. Only its nose, eyes and ears are visible.

Many aquatic animals have their nose, eyes and ears sticking up on top of their head. They can remain under water and smell, hear and see without being seen. Frogs can do this, and so can crocodiles.

Staying afloat

Staying afloat is a problem which animals solve in different ways. Oil and fat are less dense than water. Animals make use of this fact to help solve their floating problems. Fish eggs contain oil droplets which keep them afloat. Sharks have oily livers to make them more buoyant. Other fish have small balloons, called swim bladders, inside them. The swim bladders work like tiny buoys. The amount of gas inside a swim bladder can be changed to make the fish float higher or sink lower in the water. The thick fat under the skin of seals and whales helps them float, and also keeps them warm.

◀ Water plants float because they have large air spaces in their underwater stems and leaves. The trapped air makes them very buoyant. Some plants have leaves which float on the surface of water. Giant water lilies live in the Amazon River in South America. Their leaves may be 6.5 feet in diameter. Why do you think that the floating leaves have gaps in the rims? Answer on page 152.

—*Did you know?*—

A whale's nostrils open to a blowhole on top of its head.

The sailfish is the fastest swimmer. It can swim more than 60 mi/h.

Squids and octopuses swim by a kind of jet propulsion.

Sperm whales may dive to depths greater than 9,800 ft when hunting.

Fish called mudskippers are able to leave the water and come on to land. They carry their own water and air supply with them.

More about life in the water (pp.86, 87, 108-111); life in the oceans (pp.116-119); countershading (p.53)

Life in the water

Design for swimming

If you want to move, you have to push or pull against something. When you walk, your feet push against the ground and drive your body forward. The only thing animals living in water can push or pull against is the water. Aquatic animals have developed various structures for "pushing" and "pulling."

Fish "power"
A fish lashes its tail from side to side as it swims. Its tail pushes against the water and the fish moves forward. The tail fin can also change shape and work like a propeller to push the fish even faster.

Turtles, seals and whales have flippers which they use like paddles to help pull themselves forward when they swim. They can even back-paddle. Penguins have small wings which are shaped like flippers. They are too small for flight in air but penguins use their flippers to "fly" underwater.

Fish watch
Look at a fish swimming in an aquarium. Watch how it uses its tail fin and then try to work out how it uses its other fins.

Here's a clue. The different fins have different jobs to do when a fish swims.

Super diver
The sperm whale dives deeper than any other animal. Its head is full of oil which the whale uses to help it sink and float. When the whale dives, the oil becomes solid, and is called wax. Wax is heavier than oil so a diving whale is "heavy-headed." This may help it dive quickly. When it starts to come up, the whale changes the wax back to oil. Now it is "light-headed" and comes to the surface quickly.

▼ This water boatman has one pair of legs very much bigger than the others. It uses them like a pair of oars to "row" through the water.

◄ Whales swim in a different way than fish do. They have powerful muscles above and below the backbone. A whale's tail doesn't bend from side to side like the tail of a fish. Instead it moves up and down. Its flattened tail fin pushes against the water, and this pushes the whale forward.

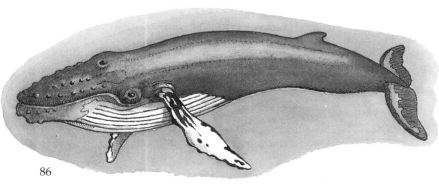

Breathing in water

Animals living in water have two ways of getting oxygen. They can either gulp it from the air above the water, or they can get it from air dissolved in the water itself. Reptiles and mammals use their lungs to breathe atmospheric air, even when they live in water. They can swim underwater for long periods because they are good at holding their breath. However, they always return to the surface when they want more air. Crocodiles and turtles do this; so do seals and whales. Adult amphibians also use their lungs for breathing, but in the tadpole stage they use gills. These are special structures which many aquatic animals have for breathing air dissolved in the water. All fish and many invertebrates living in water breathe with gills.

Built-in snorkels

Mosquito larvae have tiny trumpet-shaped structures which they push above the surface of the water. The structures work like little breathing tubes or snorkels. Some tropical snails also get their air supply in this way.

On the surface

Some animals are able to live on the surface of ponds and lakes. Surface animals are very small and have a small mass. They are so light that their weight doesn't break the surface tension of the water. It is a bit like walking over the skin on a bowl of custard. Animals like pond skaters spread their weight by standing on legs spread wide apart.

◄ A water spider's abdomen is covered with tiny hairs which trap air when it comes to the surface to breathe. When the spider has collected its new air supply, it dives down to its little "diving bell" anchored below the surface. Here it releases the air bubble into its underwater home. The water spider can even carry bubbles of air on its legs when it goes hunting under the water.

The monstrous coconut robber

The robber crab is a huge crab which lives on islands in the Indian and Pacific Oceans. Like all crabs it has gills. However, it is rather a special crab because it lives most of its life out of water. Coconuts are its favorite food. It even climbs coconut palms to get a tasty dinner. The robber crab can live out of water because it carries its own air supply. The crab keeps its gills soaked in sea water. This water contains the dissolved oxygen that the crab needs. It has to be renewed by washing the gills in sea water every day.

Did you know?

A 110 lb human would need feet with a total perimeter of 4.3 mi in order not to break the surface tension when standing on the surface of a pond.

A bull sperm whale can hold its breath for nearly two hours in a deep dive.

The Indian climbing perch is a fish which can live for long periods out of water. It can even walk over land on its front fins.

More about life under water (pp. 110, 111); life in the sea (pp. 116, 117)

Life in the air

Floating in air

Many living things are able to float in air. Very small things can float because they have a small mass. They also have a large surface area compared to their volume. Air currents keep them airborne.

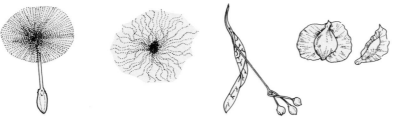

◄ These seeds have hairs and enlarged surfaces to help them float in air. These structures increase the seeds' surface area and this gives them greater resistance to falling in air.

Flying and gliding

As organisms increase in size, they develop special structures to help them stay airborne. Many seeds have hairy parachutes and wings. Larger organisms face an even bigger problem of keeping airborne. Animals have developed ways of flying. There are two types of flight. These are powered flight and gliding flight.

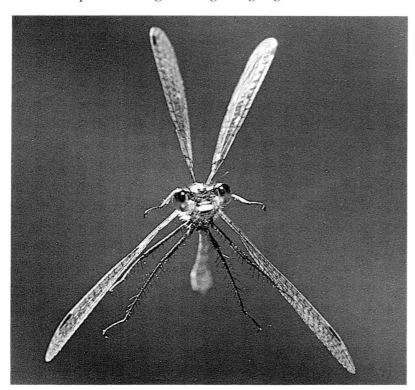

Aerial plankton

Aerial plankton contains bacterial and fungal spores, pollen grains, small animals, such as tiny insects, and even minute spiders. It is found up to many thousands of feet above the Earth's surface.

◄ Powered flight is found in insects, birds and one type of mammal, bats. The wings are formed in different ways, but they have the same function. They give the animals an increased surface area which helps them to resist falling through the air.

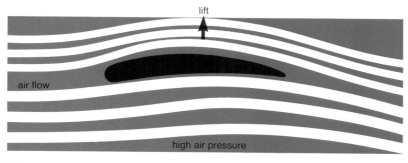

Caption: lift / air flow / high air pressure

◀ A wing cut in section looks like this. Such a shape is called an airfoil. The airflow above is different from the airflow underneath. The air pressure above the curved wing is lower than the pressure underneath, so air under the wing pushes upward and gives lift. This keeps the bird airborne. The bird's muscles provide power for flight.

The airfoil

Wings have a large surface. They are also moved by large flight muscles. They have a special shape and this is important in helping the wings gain lift. A bird's wing is curved on its upper surface, rather like in the diagram above. This curvature is important in controlling the movement of air as it passes over the wing. As air passes over the curved upper surface of a wing it has to move faster because it has farther to go. This reduces the air pressure above the wing. The air pressure underneath the wing is higher than that above. This gives lift because the air pressing up under the wing pushes the wing upward. A wing with a curved upper surface is called an *airfoil*. A bird's wings are shaped like airfoils. So are the wings of a jumbo jet. This airfoil structure, together with the bird's muscles, keep a bird airborne. The muscles provide the power for flight.

▼ Gliding flight is found in a number of different types of animal. These are not true fliers but they can glide downward from a higher place. All these animals have developed webs and flaps of skin to increase their body surfaces. This helps them resist falling through air, rather like a parachute.

A is a flying gecko. Its flattened tail helps it control its glide. **B** is a flying draco lizard. Its flaps of skin are supported by ribs and can be folded up when not in use. **C** is a flying squirrel. **D** is a flying frog. It has big membranes between its toes to increase its surface area. There is also a flying snake.

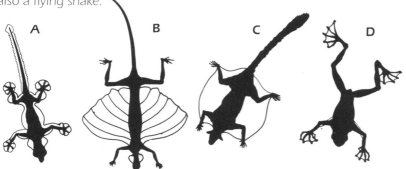

A   B   C   D

-Did you know?-

A dandelion seed may be carried 6 mi or more by the wind before landing on the soil.

Small spiders are often carried a thousand or more miles by the wind on their silken parachutes.

A bumblebee's body is covered with a "hairlike" material. This keeps the bee warm and helps its large flight muscles work more efficiently.

The paradise tree snake from southeast Asia turns its body into the shape of an airfoil when it glides down from a high tree.

A flying fish can travel or glide more than 300 ft across the surface of the sea.

More about plankton (p.117, 122); how birds fly (pp. 90, 91); size and shape (pp.26, 27)

Masters of the air

Bird adaptation to flight

Birds are the most skillful fliers in the animal kingdom. They fly in different ways but they all have similar adaptations to help them. They have well-developed wings providing a large surface area. The wings are shaped like airfoils. The bird's body is covered with feathers. These

The feathers keep the bird warm. They are also important for flight.

help increase the surface area, and also keep the bird warm. It is important that the flight muscles are kept warm if they are to work efficiently. The bird's breast muscles are large and powerful to move the wings. Inside the body the bird's skeleton is specially modified for flight. The bones are hollow. This reduces weight. The lower part of the keel bone is large and flat. The powerful breast muscles are attached to it. A bird has air sacs inside its body. These help it to breathe more efficiently. They also reduce the bird's body weight. Birds have well-developed senses, including good eyesight. This is very important for fast flight.

Flapping flight

Although some birds are skilled gliders and soarers, most birds fly by flapping their wings up and down, using their powerful flight muscles. In fast flight the muscles pull the wings downward and forward. The movement of the wing downward is called the downstroke. This movement pushes air backward and the bird moves forward. The movement of the wing upward is called the recovery stroke. When the wings are pulled up, the feathers at the tips of the wings become parted. This can be seen in the pictures below. The parted feathers allow air to pass easily between them, and so there is less resistance to air as the wings move up to begin the downstroke again.

▲ The albatross is an expert glider. It has long, narrow wings to catch the wind above the surface of the sea.

downstroke

recovery stroke

▲ The condor has very long, wide wings. It soars through the sky on currents of air.

Gliding and soaring

Some birds are experts at soaring and gliding. Vultures, buzzards and eagles have broad wings which help them to make use of warm air currents rising from the Earth's surface. These birds are soarers. Their wings spread out over a large surface. This allows them to sink slowly in air. If they find air rising faster than their sinking speed, they will gain height. The slotted wings help the warm air to flow under the wings and out through the ends. They also help the bird to turn quickly.

Birds like albatrosses, fulmars and frigate birds have long, narrow wings which make them excellent gliders. These birds catch the high wind speeds above the surface of the sea. They make semicircular turns, climbing into the wind and then descending downwind. They don't have slotted feathers at the wing tips because there is no need to turn quickly like soarers.

Hovering flight

Some birds are able to hover in the air. This is achieved by very fast wing beats. Hummingbirds from South America and their African relatives, the sunbirds, are experts at this kind of flying. Hummingbirds are the smallest birds. They use their hovering flight to "stand still" in the air while sipping nectar from tropical flowers. They can move up and down and even backwards in some cases. Because of their very small size and great use of energy, hummingbirds are not able to remain fully active for long periods. They go into a state which is rather like hibernation for about twelve hours every day.

Did you know?

A hummingbird beats its wings as many as 50 times per second.

Albatrosses on remote islands need a clear runway which they use to take off into the wind.

The wandering albatross has a wingspan of almost 10 ft.

The Andean condor weighs as much as 22 lb. It can travel more than 60 mi without flapping its wings.

More about life in the air (pp.88, 89); long distance fliers (p.78); hibernation (pp.76, 77)

Life on land

Living on land

The first living creatures evolved in the sea. Then, about three hundred and fifty million years ago, some animals moved onto dry land. Water provides a good support for the bodies of plants and animals. Think how easy it is to float in a swimming pool. But air does not provide such a good support. Terrestrial animals and plants (ones which live on land) need their own built-in supports.

Four legs and more

Worms and caterpillars move very slowly because they cannot lift their bodies off the ground. Most terrestrial animals have legs to lift them off the ground and so increase their speed. Usually they have at least two pairs of legs, making it easy to balance. Frogs, salamanders, lizards and crocodiles have legs at the sides of their bodies. They cannot lift their bodies far off the ground, so they are not very efficient walkers. Mammals' legs are attached further underneath the body, giving better support. Insects and crustaceans have rather short legs, but they are not very heavy and can easily lift their bodies clear of the ground.

▲ No legs, no speed!

Bags of water

If you fill a large plastic bag with water and tie the top very tightly, you can sit on the bag. The water inside keeps it stiff. Plant cells are like lots of tiny bags of water. The water keeps the plant stiff. If a plant loses a lot of water, for example, on a hot day, it wilts. The leaves droop because they no longer have enough water to keep them stiff. Large plants need extra support. Trees have stiff wood to support them.

Waterproofing

Living on land with your body exposed to the air means that you may dry out. Land animals and plants usually have waterproof coats. Insects and crustaceans are covered in a hard, waxy layer, called the cuticle. The skin of mammals is also waterproof. Birds coat their feathers with waterproofing oil from special glands. Reptiles are covered in scales, and are so well waterproofed that many can live in hot, dry deserts and hunt in the heat of the day. Only frogs have a moist skin which easily dries out. Frogs are usually found only in damp places.

◄ Some monkeys use their tail like an extra limb. They curl it around branches for support, and even hang by it while they feed.

▲ Both the beetle and the leaf are covered in a waterproof cuticle.

How big are your feet?

Big feet make it easy to balance, but if you want to run fast, they slow you down. Cheetahs and other cats which can run very fast walk only on the tips of their toes. Antelopes, horses and other grazing animals that travel large distances in search of food have feet reduced to hooves.

—*Did you know?*—

Millipedes don't have 1,000 legs. The largest number of legs recorded is only 355 pairs (710 legs).

Four legs are better than two! The cheetah can reach speeds of more than 70 mi/h. The ostrich can run at almost 31 mi/h, and the kangaroo at 37 mi/h.

The heaviest land animal is the elephant, at 6.6 tons. But the heaviest plant, the giant sequoia or "big tree," weighs 2,400 tons, about 360 times more.

More about life on land (pp.94, 95); moving through the treetops (p.131); waterproofing (p.22)

Life on land

Key
1. Ground beetle   2. Millipede
3. Centipede   4. Fly pupae   5. White
lipped and banded snail   6. Shrew
7. Insect eggs   8. Slug in leaf litter
9. Rabbit   10. Ant with pupae   11. Crane
fly larva   12. Beetle   13. Mole   14. Spider

The soil

Without the soil there would be very little life on land. Soil comes from tiny particles worn away from solid rocks. Around each particle is a thin layer of water. Between the particles there are air spaces. Minerals from the soil particles dissolve in the water and are taken up by plants. Soil also contains decaying organic material, or humus, from dead plants and animals.

Millions of small animals live in the soil — millipedes, centipedes, worms, beetles, ants and other insects. There is plenty for them to feed on. Hundreds of seeds and spores, as well as insect eggs and pupae, lie buried in the soil. Some soil animals feed on plant roots. Some feed on other soil animals. Many help to break down dead organic material, so releasing the trapped minerals.

Bigger animals also live in the soil. Foxes, badgers and rabbits dig deep burrows in the soil, and mice and voles tunnel through it, safe from predators. Moles spend their whole lives in underground tunnels, hunting for worms. A little way below the surface, the soil temperature stays much the same all year round. Underground homes provide warmth, shelter, and protection from bad weather.

The go-betweens

Flowering plants do not produce sperms. Instead, they produce pollen. In some plants, the pollen is carried from flower to flower by the wind. In other plants, it is carried by insects like honeybees. Many flowers contain a sugary liquid, called nectar, which insects like to drink. Colorful petals advertise the flower to passing insects. Sometimes there are patterns of lines and spots on the petals to lead the insects to the nectar. Each flower is designed so that insects searching for nectar will rub against the pollen-producing parts.

► The honeybee acts as a go-between for the flowers.

▼ Pollen from hazel catkins blows away on the breeze.

▲ When they reproduce, most animals living in water shed their sperms and eggs into the water. The sperms swim to the eggs. Except in damp places, there is not enough water for this to happen on land. Most land animals pass their sperms directly from the male into the female, who keeps the eggs inside her body until they have been fertilized and sometimes even longer.

Waterproof eggs

Most land animals lay eggs which have tough, water-proof coats. Only animals living in damp places have soft, moist eggs. Snails and slugs lay their eggs in the soil, which stops the eggs from drying out. Some animals get around the problem by keeping the eggs inside them until they hatch. The young are born fully developed. Mammals do this, and so do some snakes and lizards.

— Did you know? —

One gram of soil may contain as many as 4000 million bacteria.

Termites use soil to build huge mounds up to 6 m high. Some mounds have their own central heating system, using decaying vegetation to generate heat. Others have a complicated system of chambers and passages which form an air-conditioning system.

More about life on land (pp. 92, 93); feeding on nectar (p. 33); how animals reproduce (pp. 66, 67)

Life without water

Deserts of the world

Deserts are usually found in areas which get very little rain, or where rain falls rarely, perhaps only once every seven years. In hot parts of the world, much of the rain falling on the scorching ground evaporates before it has time to sink into the soil.

Some deserts are sandy. They have rows of sand dunes that move with the wind. Others are bare and rocky, with high mountains and deep, dry canyons. Deserts are places of extremes, with blazing hot Sun by day, and very cold temperatures by night, and long periods of drought followed by sudden torrential rain.

The long sleep

Many plants and animals avoid drought altogether by sleeping for months at a time. This sleep is a kind of hibernation called estivation. The dry season is also a time when food is scarce, so estivation saves food as well as water. The African lungfish can survive for four years in the dried-up mud, breathing through a narrow tunnel from its burrow to the surface. The desert tortoise sleeps in a deep hole underground, storing water in its bladder.

Plants need water too. They rely on water in their tissues to keep them stiff. Without it they wilt. Many desert plants spend most of their lives as dormant seeds, or as underground bulbs and stems swollen with stored food. The soil protects them from the Sun's drying rays. When rain falls, their stored food is used to produce leaves, flowers and seeds quickly, while the water supply lasts.

◀ Even in rocky deserts some plants manage to survive.

◀ The spadefoot toad of North America uses a special horny pad on its hind feet for digging. It digs down as much as 10 feet below the desert surface to escape the heat of the dry season.

▼ The regal-horned lizard hunts in the early morning and later afternoon, when it is warm enough for a cold-blooded animal to be active, but not too hot for comfort.

Fish in the desert

After heavy rain, pools and lakes appear in the desert. Soon they are full of life — shrimps, water fleas, fish and even tadpoles. But where have all these creatures come from? Many of them have spent the dry period resting as eggs or pupae in the soil, protected by waterproof coats. Eggs and pupae do not move or grow, so they use very little stored food and need very little water.

Like many desert plants, the animals of these temporary pools pass through all the stages of their life cycle in a surprisingly short time. The African bullfrog tadpole hatches from its egg less than a day after rain falls. In just four weeks it is an adult frog. After mating and laying eggs, it buries itself in the mud, protected from drying out by a cocoon of mucus and mud. Here it sleeps until the next rains.

—*Did you know?*—

Fairy shrimps survive the desert drought as tough-coated eggs. They can survive for 100 years without water, and still hatch successfully after rain.

The hottest land temperature ever recorded was 136°F in the Sahara Desert.

Desert frogs and toads can store up to 40 percent of their body weight as water in their bladders.

A food store in the soil! The soil of the Californian desert receives about 1½ billion seeds a year, more than enough for all the animals that would like to eat them.

In the Arctic and Antarctic there is very little rain — only snow falls. These regions are cold deserts.

More about life without water (pp. 98, 99); hibernation (pp. 76, 77); eggs and pupae (pp. 68, 69)

Life without water

Cacti, the desert specialists

The giant saguaro cactus lives for up to two hundred years. It grows to a height of fifty-six feet, and weighs about eleven tons. Four-fifths of this weight is water. Cacti like the saguaro are well adapted to the desert. They store vast amounts of water in their swollen stems. The stems are ribbed, like a concertina, so they can expand when full of water, and shrink as they dry out. Cactus roots are not very deep, but they spread sideways, for some distance, to trap rainwater. Their stems are green, and they make their own food by photosynthesis. They have thick, waxy, waterproof surfaces, and spines which keep off browsing animals.

Living through the drought

Some desert animals stay active throughout the year. Birds, and large mammals like the desert fox, the oryx and the kangaroo, can travel to find water. Many animals spend the hottest part of the day in burrows, only coming out to feed in the cool of the evening. The air inside the burrow becomes damp from the moist air that the animal breathes out. The kangaroo rat plugs the entrance to its burrow to keep in the moisture.

Weaver birds in the African deserts live in huge communal nests woven from dry grasses. A weaver bird colony looks rather like an untidy haystack. Inside the nest, the birds are protected from the Sun, and the air stays cool and moist.

Water-hoarding plants

Living in the desert is difficult for many plants. Some, like the tamarisk, have deep roots which reach permanent water supplies far below the ground. Others, like cacti, have shallow roots which are spread over a wide area to catch as much rain as possible before it evaporates. Many desert plants store water in their stems or leaves. Only a few plants, such as the strange resurrection plant, can survive being completely dried out. Some desert shrubs, like the ocotillo, even shed their leaves in dry weather so as not to lose water from them.

Plants are often the only source of moisture for desert animals. Many desert plants have spines to keep thirsty animals at bay. The stone plants, *Lithops*, rely on camouflage instead. They have only two leaves, which look like the pebbles which surround them.

▲ A kangaroo rat.

Lord of the desert

Everything about the camel is adapted to life in the desert. Its hump is full of fat, a store of food for the long dry season. The hump shields its back from the Sun. It acts like a large sun hat. When the camel is resting, the hump insulates it from the heat of the Sun, and keeps the body cool. The rest of the body has very little fat, and can easily lose heat from its surface. The camel can drink seventy quarts of water at one time, one-third of its own body weight. It even swallows its own nose drippings to save water. The camel's feces are so dry that they can be used to light fires. Its huge feet are wide and spreading to prevent the camel from sinking into the sand.

Cooling down

Some desert animals have large ears. As blood runs through the ears, it loses heat to the surrounding air, and cools down the animal.

Drinking the dew

There is another source of water in the deserts. As deserts get cold at night, moisture in the air condenses as dew. In the early morning, animals lick the droplets off the plants, or simply eat the wet leaves. In some deserts, such as the Namib Desert of Southwest Africa, fog is common. Here, desert beetles stand on their heads, letting the fog condense on their bodies and trickle down into their mouths.

More about life without water (pp. 96, 97); camouflage (pp. 52, 53); photosynthesis (pp. 28, 29)

Life in cold climates

The long winter

In the Arctic and Antarctic it is cold throughout the year. There is seldom any rain — only snow falls. The winters are long. In midwinter the Sun never appears above the horizon. In summer the days are long, and the Sun does not set. The summer Sun is low in the sky, and does not give much warmth. Animals that live in these areas cope with these freezing conditions all year around.

Farther away from the polar regions the weather is warmer, but the winters are still cold and snowy. There is a bigger difference between the seasons, and the summers are quite warm. Animals living here have a choice. Either they can hunt for food in the cold of winter, or they can avoid the worst of the weather by hibernating or migrating.

Growing your own blanket

Mammals that live in cold regions usually have thick fur. Often they grow extra-thick fur in the autumn, and moult to a thinner coat in the spring. The musk ox has some of the thickest fur. It grows an extra layer of soft, warm fur under the straggly, waterproof, outer layer. It looks as if it is draped in a shaggy old rug.

▲ Many animals put on extra fat during the late summer. This provides a food store for the winter. Fat is also a good insulator (it does not allow heat to escape). The extra fat helps to keep the animal warm. Mammals like seals and walruses, which live in the polar seas, have thick layers of fat rather than extra fur.

Snow can keep you warm

Small animals lose heat easily. Many birds and small mammals use the snow as an insulator. The ptarmigan scrapes a hollow in the snow to shelter from the icy winter wind. Lemmings tunnel under the snow, feeding on roots and seeds in the soil under the warm, white, snowy blanket.

Lying low

In the far north, the bitter, cold winds and lack of rain prevent trees from growing. Only small, slow-growing plants survive. This treeless region is called the tundra. It is covered in lichens, mosses and cushion plants. These plants are low and rounded. Because they are so short, they can avoid the worst of the wind, and are protected under the snow in the winter.

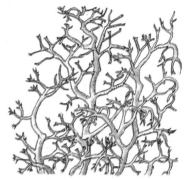

▲ There are lots of feathery lichens. These are called reindeer moss, because they are the favorite food of the reindeer.

Born under the ice

The polar bear gives birth to her cubs in the middle of winter in a den inside the ice of the frozen polar sea. She remains with them, living on her store of fat, until they are big enough to explore the world outside. By this time spring has arrived, and there is plenty of food around.

Did you know?

In polar regions, insect eggs and pupae contain "antifreeze" to help them survive the winter. So do plants.

Fish living in the polar seas have supercooled blood. Although the blood temperature can be below freezing point, "antifreeze" in the blood prevents it from freezing.

Many Arctic seabirds have very cold feet. This prevents them from losing heat from their feet, which are usually not protected with feathers.

Blood runs very near the body surface in ears. The Arctic fox has much smaller ears than any of its relatives. The tiny ears reduce heat loss.

More about life in cold climates (pp.102, 103); hibernation (pp. 76, 77); migration (pp.78, 79)

Life in cold climates

▲ In summer the ptarmigan has a dull, speckled coat that blends with the summer plants.

▼ By the winter the ptarmigan's coat is white to match the snowy landscape.

▲ Snowshoe hare.

The winter hunters

Animals that hunt in the snow all year round — the polar bear, Arctic fox and snowy owl — have permanent white camouflage. Many animals that feed on the tundra in summer change the color of their coats with the spring and autumn moults. Ptarmigan blend perfectly with the summer tundra, but change to winter white in the autumn. So do hares, foxes and stoats. The stoat's color change is triggered by falling temperatures. Farther south stoats turn white only in very severe winters.

Snowshoe feet

Winter hunters need to be able to walk on snow. The snowshoe hare has huge feet. They spread its weight over the snow, so it is less likely to sink down.

Reindeer and caribou also have large spreading feet. These help in spring, too, when the thawing snow makes the ground soft and boggy.

The ptarmigan grows its own snowshoes. In winter, it has extra feathers on its feet. These keep the feet warm as well as acting like snowshoes.

The snow leopard lives in Asia. It has thick hairs between its toes. These also act like snowshoes.

The polar spring

The insects of polar regions pass the winter protected in the soil as eggs or pupae. When spring comes, they emerge in their millions to feed on the new growth, and on the nectar in the tundra flowers. The melting snow produces a land-scape full of pools, and mosquitoes thrive. They become such a nuisance that the caribou will sometimes go and stand in the sea to get away from them.

This wealth of food makes the tundra an attractive breeding ground for birds. Thousands of birds fly north, often traveling thousands of miles to rear their young in the tundra. The long summer days allow plenty of time to find food to feed the growing families.

In early spring the weather is still very cold. But the birds must lay their eggs well before all the insects hatch, as the summer season is so short. Eider ducks line their nests with soft, downy feathers.

▲ Penguins incubate their single large egg on their feet. Their legs are very short, and they have a special hollow in their feathers which fits snugly over the egg.

▲ Cranes fly south to avoid the winter.

Running away

Many animals avoid the polar winter by migrating to warmer areas. The North American caribou (reindeer) feed on the tundra in summer. In autumn they migrate 435 miles south to the forests. Here, the snow, although deep, is less frozen, and they scrape it away to reach the plants below. They are followed by packs of wolves and by scavengers.

Penguins in Antarctica spend the winter at sea. They may travel 60 miles back to their breeding sites in spring, partly swimming and partly walking across the pack ice.

Did you know?

The polar bear is perfectly camouflaged except for the tip of its nose. When it is stalking prey, it will cover its black nose with a paw.

Arctic poppies can turn their heads to follow the Sun as it moves across the sky. This keeps the centers of the flowers warmer than the surrounding air, helping to attract insects.

Seals spend the winter hunting under the ice. They make their own breathing holes in the ice.

More about life in cold climates (pp.100, 101); camouflage (pp. 52, 53); migration (pp.78, 79)

103

Grasslands

▲ The pronghorn can run faster than any other American mammal.

The great plains

Stretching across the centers of most of the continents are vast areas of grassland, including the steppes of Europe and Central Asia, the savannas of Africa and Australia, the prairies of North America, and the pampas of South America. In some areas the grasses grow to over six feet tall; in others they are kept short by herds of grazing animals. Once there were millions of these grazing animals, but they have been hunted so much that their numbers are now much smaller. In Africa there are still large herds of antelopes and zebra, but the great herds of American bison have gone, the European bison are near extinct (there are some in captivity), and the wild horses of Asia are rare.

▶ In Australia, kangaroos graze on the savanna. They can leap across the grass-lands at a great rate in search of better feeding areas.

Long-distance travelers

The herds of animals on the grasslands roam far and wide in search of good grazing. Often they migrate with the seasons, following the rain. They are not the only long-distance travelers. Ostriches (Africa), rheas (South America) and emus (Australia) also feed on grassland plants. They are large birds and cannot fly, but they have very long, powerful legs and can run very fast.

Life underground

The plains offer little shelter for small animals. Most live in underground burrows, which protect them from the heat of the day, and from the cold in winter. They also provide safe havens from predators and fire.

For many of these animals danger comes from both above and below. Snakes can slide into burrows unnoticed. Eagles and other birds of prey scan the plains from the air, alert to every small movement in the grass. Members of the dog family — foxes, coyotes, hunting dogs and dingoes — are common predators in grasslands. In Africa, big cats such as lions and cheetahs also prowl the plains, and in South America the puma stalks through the pampas. Vultures are a common sight, circling high above the grasslands in search of carcasses.

Builders of the plains

Large termite mounds are dotted across the grasslands. Ant- and termite-eating animals are common in the grasslands. For example, the giant ant-eater of the pampas, the aardvark and aardwolf of the African savanna, and the spiny anteater of Australia. They all use strong claws to tear into the ant or termite nest, and long sticky tongues to extract their prey.

Food for all

Grasslands are full of food for small mammals. Grasses and other flowering plants produce millions of seeds. The underground parts of plants provide more food, and the insects attracted to plants can be eaten, too. Grasslands are home to many different kinds of mice, rats and voles, and, in Europe and Asia, hamsters and gerbils. Many of them collect huge amounts of seeds and store them in their burrows for the winter or the dry season.

Fire makes the grass grow

Many grassland plants flower only after a fire. Most of them rest underground while the grass is high, surviving as bulbs or underground roots and stems swollen with stored food. Fire is common on the plains. The thunderstorms that end the dry season produce lightning, which easily ignites the dry grasses. After the fire and rain, tender new shoots of grass appear, providing better fodder for the grazing animals. The plains are now full of color from lilies, tulips, irises, anemones, gladioli and many other flowers. They have to produce their seeds before the grass grows tall and shades them again.

Did you know?

The mounds of some termites can be over 40 ft high.

Although they have small brains, ostriches are quite intelligent, and can even be trained to herd sheep.

◄ In America, prairie dogs live in vast underground towns, building entrance mounds which serve as lookout posts.

More about anteaters (p.39); migration (pp.78, 79); animals that eat plants (pp.32, 33)

105

Temperate forests

▲ Spotted deer in English forest.

Forests of the temperate regions

Temperate regions are parts of the world which have warm summers and cool or cold winters. At one time, large areas of the world's temperate regions were covered in forests of broad-leaved trees. These forests are mostly of deciduous trees, that is, trees which lose their leaves in winter. Many forests have been cleared for farming. The remaining forests are home to many different animals.

Vegetarians of the forest

In the summer there is plenty of food for plant-eaters. The caterpillars of many moths feed on the leaves. Mice and voles make their homes among the twisting roots, or burrow in the soft soil. They eat tender young buds and seeds. The American porcupine eats the bark as well. The sapsuckers of American forests are birds which look like woodpeckers. They drill holes in the bark of trees to get to the sap below. At dawn and dusk, deer browse on young shoots, and rabbits scamper across the forest clearings. In autumn, nuts and berries add to the feast, just as the leaves are beginning to fall.

▶ Once the leaves have come out in the spring, the forest floor does not get much light. Many forest herbs put out leaves and flowers before the trees come into leaf. They can do this because they have underground stores of food in swollen roots and stems, or in bulbs and corms. Snowdrops, anemones, daffodils, celandines, bluebells, and, in America, trilliums and hepaticas, form carpets of color in the spring.

Hunters of the forest

Many predators hunt the plant-eaters. Shrews search for insects on the forest floor, and many small birds such as tits, warblers, wrens and thrushes feed on the caterpillars. There are not many wolves and mountain lions left, but there are plenty of foxes, weasels, wildcats and martens, as well as hawks and owls. After dark, badgers and hedgehogs hunt for small invertebrates, and bats chase night-flying insects through the trees.

Every forest has its all-rounder. In America it is the raccoon which feeds on almost anything — insects, mice, frogs, and even fruit and nuts. In Europe, wild boars root for nuts and fungi, and anything small enough to be caught without much effort.

◀ Young red fox.

▶ Woodpeckers use their strong beaks to break open bark and rotten wood. They use their long sticky tongues to extract the insects and spiders living there.

The winter woodland

In winter the leaves are gone, and it is not so easy to hide from predators. Animals which feed on leaves have very little food now, and soon even the nuts and berries will all have been eaten. Insect-eating woodland birds like the warblers migrate to warmer climates, where food is plentiful.

Some animals store food for the winter. Squirrels bury nuts in the woodland floor, and so do jays. Mice carry nuts and seeds off to their nests.

The insects spend the winter as eggs or pupae, buried in the soil or hidden in the crevices of the tree bark. Small animals like mice, voles and squirrels spend a lot of the winter sleeping in their warm nests. Some, like the dormouse and the chipmunk, actually hibernate during the coldest months. So do bats, which huddle together for warmth. Even the black bear of North America goes into a very deep sleep, waking up occasionally to search for food.

Did you know?

The acorn woodpecker of North America drills holes in old tree stumps and fills them with acorns, which it eats later when food is scarce.

More about forests (pp. 10, 11); animals that eat plants (pp. 32, 33); hibernation (pp. 76, 77)

Life in rivers and streams

Swimming against the current

Animals which live in rivers and streams are good swimmers. Most mammals push with their feet when they swim. Mammals like water voles and shrews, otters and mink, which regularly hunt in rivers and streams, have webbed hind feet to help them swim. Webbed feet also help them walk on the marshy ground beside the river.

Why is an otter like a crocodile?

The otter is especially well equipped for life in the water. Its fur is oily and waterproof. The otter's body is smooth and streamlined, and its ears are small. You might not think that an otter is like a crocodile, but they both have webbed feet, a flattened tail for pushing against the water when swimming, and ears and nostrils that can be closed when under the water.

Lazy grazers

In several parts of the tropics, the lush vegetation of large, slow-flowing rivers is grazed by manatees or dugongs, "sea cows." These large docile mammals swim slowly using flippers. They have smooth fat bodies rather like seals. They tear at the plants with their long muscular lips. Sadly, their numbers are low today because they are so easily caught for food.

Fishermen large and small

Many of the hunters in the world's rivers are very large indeed. Crocodiles and alligators are common in the tropics, where they bask in the Sun. They open their huge mouths to cool themselves, showing their long rows of teeth. River dolphins also have long rows of pointed teeth, over one hundred in all, for catching fish.

South American piranhas make up for their small size by their numbers. They hunt in groups of hundreds or even thousands. Mostly they catch other fish, and occasionally animals like capybaras (the biggest rodents in the world). They also clean up carcasses that fall into the water.

Danger on all sides

There are predators on the river banks, too. Kingfishers scan the water from overhanging branches, then dive directly on their prey. The heron stalks its prey through the shallower water. The purple gallinule, an American bird rather like a moorhen, lures prey into its mouth by wiggling its wormlike tongue.

Some river turtles eat crustaceans and even small fish. The matamata turtle can stay under water for long periods. It has a very long neck and its nose acts like a snorkel, taking in air from the surface. It catches the prey by waiting for it to come within reach. Then the turtle suddenly opens its huge mouth and sucks in the prey.

Holding on tight

In fast-flowing streams it is easy to be swept away. The eel-shaped river lamprey has a large, suckerlike mouth which helps it cling to rocks. The river lamprey can even hang on to rocks under waterfalls. The climbing catfish of South America also has a suckerlike mouth. Its belly fins have tiny toothlike structures which stop the catfish from slipping.

▲ A river lamprey clinging to a rock in a fast-flowing stream.

The world's biggest rodents

The plants of South American river banks are food for the capybara, the largest rodent in the world. An adult female may be up to 4.3 ft long, and can weigh 120 lb. Capybaras have long legs with webbed hind feet, and are very good swimmers.

—Did you know?—

South American hatchet-fish leap out of the water to catch flying insects.

The largest freshwater fish in the world is the South American arapaima. It can grow to over 10 ft long and can weigh 300 lb.

In the Amazon there is a fish with four eyes. It lives near the surface of the water, and has one pair of eyes for seeing under the water and one pair for seeing in the air above.

More about swimming (p.86); streamlining (p.84); life in the water (pp.110, 111)

Ponds, lakes and swamps

Life at the surface

Ponds, lakes and swamps have lots of places for animals to live. Some live right at the water's surface. Fishing spiders lie in wait on floating leaves, or on the surface of the water, trailing a foot on the surface film. They can sense the vibrations from small fish near the surface, and know just when to pounce.

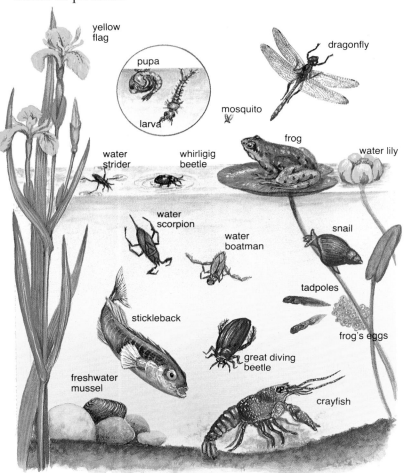

yellow flag

pupa

larva

dragonfly

mosquito

frog

water lily

water strider

whirligig beetle

water scorpion

water boatman

snail

tadpoles

stickleback

great diving beetle

frog's eggs

freshwater mussel

crayfish

Life among the water weeds

The underwater plants are food for many small animals. Frog tadpoles graze on the algae coating the underwater leaves and stems. Pond snails glide over the leaves, depositing jellylike masses of eggs behind them. Dragonfly nymphs and water beetle larvae lie in wait for other small water insects, tadpoles, and even tiny fish.

The big hunters

The edges of lakes attract many animals. While they are drinking they are easily caught by surprise. In tropical waters, crocodiles and alligators lie in wait just below the surface of the water, with only their eyes and nostrils showing. They will attack large mammals, like deer, when they come to drink. So will the anacondas, South American snakes up to 20 ft long. Even European grass snakes and American cottonmouths are good swimmers, and can catch frogs and toads.

The lakes themselves are full of fish, which are food for larger carnivores. In deep lakes some fish can grow very big. The European pike may weigh up to 50 lb. It eats other fish, even smaller pike, as well as mammals like water rats.

Making their own lakes

Beavers make their own lakes by damming rivers with logs and branches, using their sharp teeth to fell trees. They build large lodges of branches and sticks, held together with mud. There are many entrances under the water. In these lodges the beavers bring up their young and store food for the winter. Beavers have changed the landscape of parts of North America, making large lakes and changing the flow of rivers and streams.

▲ For all their great size, hippopotamuses feed on grass. At night they leave the water and graze in the nearby water meadows, like giant lawnmowers. By day, they wallow in the water, stirring up the mud at the bottom. They pass huge amounts of dung into the water. The nutrients from the dung feed the algae which feed the fish, which feed the birds, and so on.

Danger from above

Around the edges of lakes there are often dense reed beds where birds can find safe nesting places out of sight of predators. Many birds hunt in lakes and ponds. Fish eagles swoop on their prey from above, seizing it in their talons. Fishing bats feed this way, too. Herons stalk their prey in the shallow water. Flamingos and spoonbills sieve the water through special bills to filter out crustaceans. Pelicans scoop up fish in their huge throat pouches. Ducks paddle along the surface, or dive and swim under the water, feeding on water weeds and small water creatures.

Did you know?

Giant otters weighing up to 75 lb live in South American lakes.

The deepest freshwater lake in the world is Lake Baikal, in the USSR. It is 5,315 ft deep, and contains one-fifth of the world's fresh water. Over 1,200 different animals and about 600 different plants live in the lake.

More about life at the surface (p.87); pond animals (p.69); life in the water (pp. 108, 109)

Life between the tides

The longest habitat in the world

The world has 193,000 miles of coastline. Along these coasts the tide rises and falls every day. Between the high tide mark and the low tide mark lies the intertidal zone. Here live very special groups of plants and animals. In some places the intertidal zone is only five or six feet wide, in others it may be close to two miles. Some shores are rocky, some have sandy beaches, and some are flat and muddy. They may look rather empty, with no life, but a square foot of rock in the intertidal zone may be home to 100,000 animals. You may not see many of them. Some will be too small to see; others come out only at night, or when the tide is covering them.

It's a hard life

Animals and plants which live in the intertidal zone have to cope with a lot of difficult conditions. When the tide comes in, the waves pound the shore, and the animals are at risk of being washed out to sea. At high tide they are under water, but at low tide they are exposed to the Sun and wind.

Life beneath the sand

On muddy and sandy shores, the animals can bury themselves in the sediments at low tide. Some make fairly permanent burrows. When the tide is in, they draw in water and extract food from it. Many worms and mollusks do this. Cockles and other mollusks with hinged shells send up tubes, called siphons, to the surface of the sediments to feed at high tide. When they die, their shells lie spread out on the sand and mud. Some marine worms make special protective tubes of sand and mud, and put out their feathery tentacles to feed at high tide. The ghost crab waits at the entrance to its burrow with only its stalked eyes showing. It will pounce on any small creature that comes within reach.

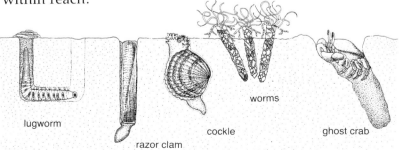

lugworm
razor clam
cockle
worms
ghost crab

Tracking snails

Choose a piece of rock in the intertidal zone. Take some enamel paint, and paint some dots on the shells of a few limpets and periwinkles. Make a map to show their position. Wait until the tide has covered the rocks and gone out again, then look for your spotted shells. Compare them with your map, and see how far they have moved.

Exploring rock pools

Choose a rock pool and see how many different creatures you can find (don't forget the seaweeds). Try turning over the stones and lifting up the seaweed. If you want to fish, a small, strong net with a wooden edge is a good idea, as you can push this under weeds.

The rock pool

1 **Lichens** Grow on bare rock. Different colors at different levels of the shore.

2 **Seaweeds** Cling to rocks with tough holdfasts. Flexible, so waves don't break them. Slimy mucus prevents them drying up. Different species occur at different levels on the shore.

3 **Sea slaters** Scavenge among rocks.

4 **Mussels** Have special threads to fix them to rocks. Close their shells to avoid drying out. Filter feeders.

5 **Barnacles** Permanent fixtures. Filter feeders.

6 **Limpets** Clamp on to rock so hard that they wear a groove in it. Graze on algae when tide in. Always return to same resting place.

Dustmen of the seashore

The remains of many ocean-dwelling animals get washed up on the beach, especially after storms. They all collect along the drift line, together with the corpses of shore animals and broken-off seaweeds. Some animals, like the seagull, find them an easy source of food. Crabs are the "dustmen" of the shore. They will eat almost anything. After sunset, the drift line is alive with sea slaters and sandhoppers, feeding among the decaying seaweeds.

7 **Periwinkles** Feed on seaweed. Shelter under weed, or in crevices at low tide.

8 **Whelks** Also shelter in crevices. Eat other animals.

9 **Crabs** Shelter under rocks or seaweed. Eat almost anything.

10 **Red coralline algae** Deposit skeletons of limestone.

11 **Anemones** Catch other animals with their stinging tentacles.

12 **Hydroids** Feed like anemones.

13 **Fan worms** Filter feeders.

14 **Sponges** Filter feeders.

15 **Sea squirts** Filter feeders.

16 **Starfish** Prey on other animals, including shellfish.

17 **Flatworms** Scavengers.

18 **Ribbon worms** Feed on other marine worms.

19 **Prawns** Filter feeders.

20 **Blennies** Eat other small animals. Camouflaged to escape attention of seagulls. Can change color.

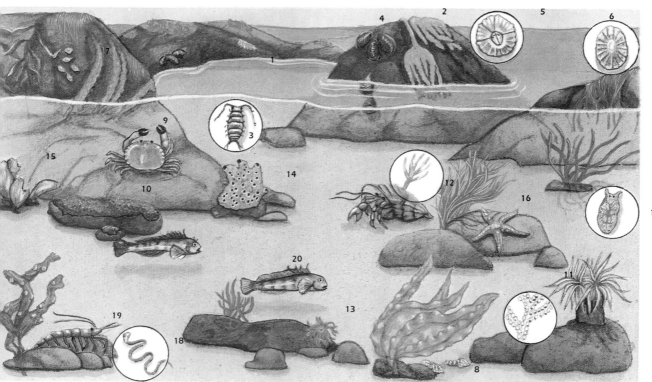

Forests on the march

▲ The tangled roots of the mangrove trees trap mud and pieces of vegetation carried to the swamps by rivers and the sea. As more and more mud collects, the mangrove forests gradually creep farther into the sea, growing on top of the new layers of mud. In this way, the mangrove swamp slowly reclaims land from the sea.

Life in a tangle

Mangrove swamps are found on sheltered tropical coastlines. They are a halfway house between the sea and the shore and are home for a fascinating group of animals and plants. They are muddy habitats which are covered twice every twenty-four hours by the tide. The main plants growing in these swamps are mangrove trees of various kinds. The trees are specially adapted for growing on mud, and they are also able to live in the very salty conditions.

One of the most common animals found in the swamps is a small fish called a mudskipper. At low tide the mudskippers flip across the mud using their fins like small crutches. One species even uses a modified fin to climb up the mangrove roots. Mudskippers can survive for long periods out of water. They carry their own oxygen supply trapped in water around their gills. This oxygen reservoir has to be renewed from time to time by dipping the gills in water. One species builds its own swimming pool on the mud and surrounds it with a wall. It defends its territory against all trespassers.

The root of the trouble

Getting a hold in a layer of moving, sticky mud is not easy. It helps to have special roots for the job. Trees at the edge of the sea produce a platform of roots, rather like a raft, just below the surface of the mud which they sit on. These are the pioneers of the swamp. They colonize new ground. They also produce roots which grow upright above the surface of the mud. These are used for "breathing." Most of the food is found in the top inch or so of the mud. A root raft just below the surface is just right for supporting the trees and for taking in food. Trees farther from the sea grow special prop roots to hold them on the mud.

▼ Mudskippers live in mangrove swamps in coastal areas of many parts of the tropics.

The "water-pistol" fish

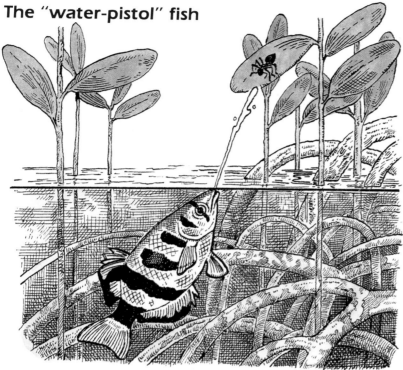

When the tide comes in, it brings with it an unusual fish with a remarkable way of catching food. It is called the archer fish, but a better name would be the "water-pistol" fish. It swims around the underwater roots of the mangrove trees using its big eyes to focus on objects above the level of the water. If it spots an insect lurking on a leaf, it squirts out a jet of water and tries to knock the insect off its perch. It doesn't always succeed, but when it does, it gobbles up the insect as soon as it falls into the water. The fish makes its aim more accurate by getting right underneath its prey before firing its water gun.

The mangroves of Borneo are home for the long-nosed or proboscis monkey. Only the male has a big nose. It probably works like an amplifier, increasing the sound of its honking among the forest trees. The monkey is an expert swimmer and often jumps into the water when crossing from one part of the forest to another.

◀ At low tide the surface of the mud becomes alive with crabs of various kinds, including the fiddler crab. The male has one front claw which is very much bigger than the other. It is also much more brightly colored. It is seldom used for fighting, but its bright colors are used for advertising, and attracting a mate and also for threatening a rival.

— *Did you know?* —

The crab-eating frog is the only amphibian which can live in salty conditions. It lives in the mangroves of southeast Asia where it feeds on crabs, scorpions and insects.

More about breathing in water (p.87); advertising colors (p.64); life between the tides (pp.112, 113)

The open ocean

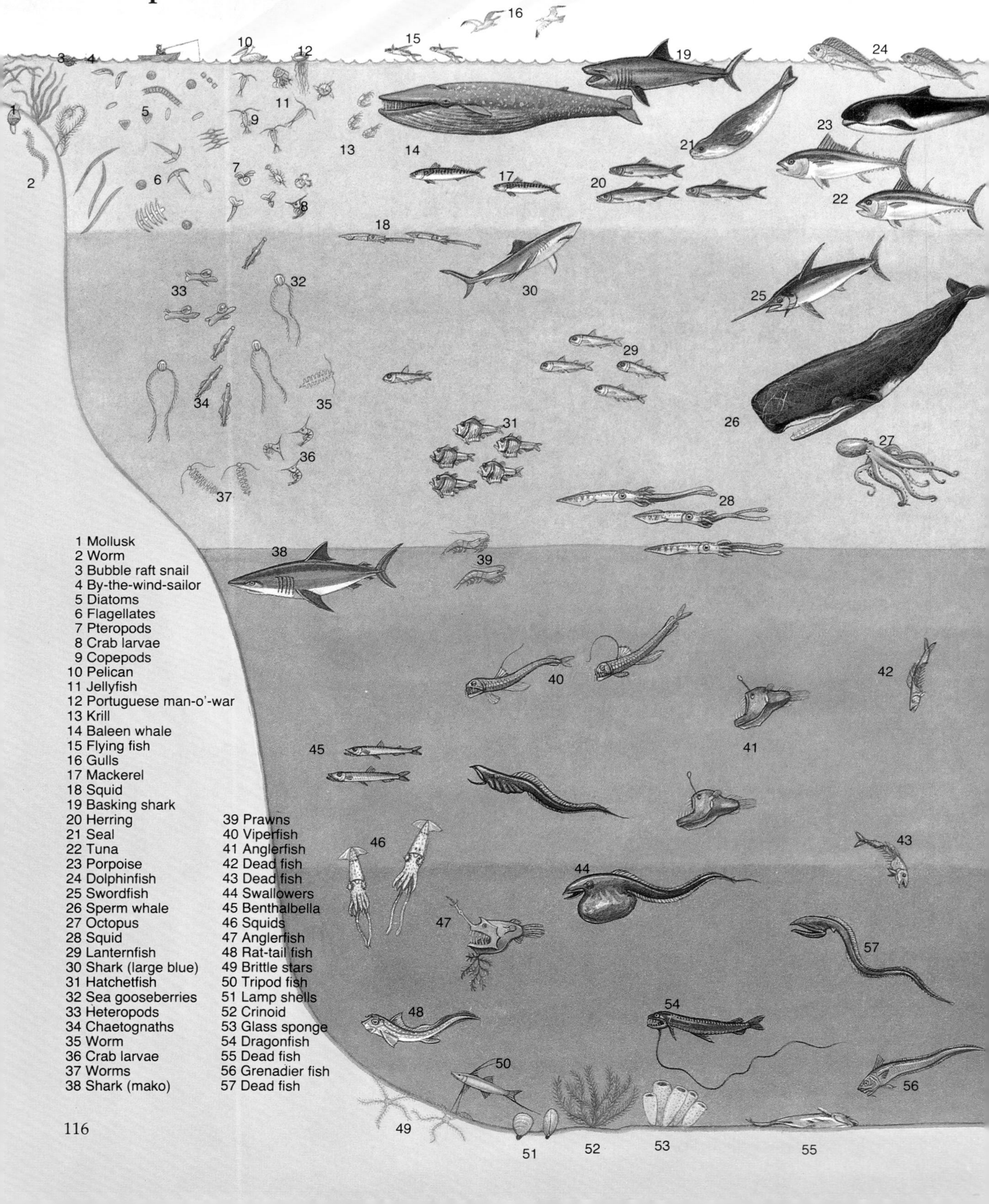

1 Mollusk
2 Worm
3 Bubble raft snail
4 By-the-wind-sailor
5 Diatoms
6 Flagellates
7 Pteropods
8 Crab larvae
9 Copepods
10 Pelican
11 Jellyfish
12 Portuguese man-o'-war
13 Krill
14 Baleen whale
15 Flying fish
16 Gulls
17 Mackerel
18 Squid
19 Basking shark
20 Herring
21 Seal
22 Tuna
23 Porpoise
24 Dolphinfish
25 Swordfish
26 Sperm whale
27 Octopus
28 Squid
29 Lanternfish
30 Shark (large blue)
31 Hatchetfish
32 Sea gooseberries
33 Heteropods
34 Chaetognaths
35 Worm
36 Crab larvae
37 Worms
38 Shark (mako)

39 Prawns
40 Viperfish
41 Anglerfish
42 Dead fish
43 Dead fish
44 Swallowers
45 Benthalbella
46 Squids
47 Anglerfish
48 Rat-tail fish
49 Brittle stars
50 Tripod fish
51 Lamp shells
52 Crinoid
53 Glass sponge
54 Dragonfish
55 Dead fish
56 Grenadier fish
57 Dead fish

The oldest environment on Earth

The open ocean is probably the oldest environment on Earth. The ocean has changed little over millions of years. Its surface area is twice the area of the land, and its average depth is two and a half miles. This gives it a volume of three hundred and twenty-seven million cubic miles, all of it occupied by living creatures.

The plankton

The surface waters of the ocean have the most animals. Here, billions of microscopic one-celled plants called diatoms, encased in glassy shells, are the main source of food and energy. In these sunlit waters lives a vast drifting community of tiny plants and animals called plankton. Most of the plankton animals are crustaceans, mainly tiny copepods, which swim using their two large antennae. There are also small shrimps and prawns, and the larvae of many other animals — starfish, sea urchins, sea snails, crabs, barnacles and a host of others. They include tiny herbivores, carnivores and filter feeders. There are so many of them that the great baleen whales eat nothing else.

Ocean surface drifters

Among the plankton are some larger animals, the ocean surface drifters. Many of them have gas-filled floats which act like sails, catching the wind and blowing the animals along. Jellyfish like the Portuguese man-o'-war and the by-the-wind-sailor drift across the sea, trailing their stinging tentacles behind them. Other drifters feed on them. The bubble raft snail has a float of bubbles, while the sea slug *Glaucus* gulps in air at the surface to help it float.

▼ Bubble raft snail eating by-the-wind-sailor.

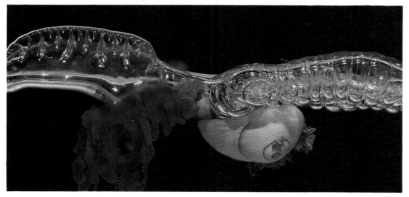

▲ Copepods are the most important animals in the plankton.

Surrounded by danger

The animals of the upper waters have enemies both from above and below. Seabirds like pelicans, boobies and gulls swoop on them from above, and seals and penguins chase them under water. Many of them are colored deep blue on their backs and white below. From above, their blue backs match the deep blue of the ocean. From below, their white bellies blend with the light coming from above.

Fish great and small

In the deeper waters of the ocean are many fish. Some, like the herring, swim along with their mouths open, filtering the water to sieve out the plankton food. Other predators chase after their prey. The flying fish has winglike fins and can glide through the air for 700 feet to escape predators in the water below.

The ocean deeps

Life in the ocean deeps

The average depth of the ocean is 2.5 miles, and parts of it reach depths of 6.8 miles. The temperature falls as the ocean deepens, until, at very great depths, it may be below 32°F. At the same time, the pressure increases because of the great weight of the water above. This makes it difficult to move quickly.

The ocean twilight

The light is dimmer in the ocean depths. Blue light goes deeper than red light, so the scenery appears bluer as the ocean deepens. Animals adapt to the changes in light by using different camouflage. Deep-sea prawns are usually bright red. There is no red light at these depths, so they will normally appear black.

Many fish living at depths between 325 feet to 6,550 feet are dark on top, but have silvery mirrorlike sides. By day, the sides match the light around them, and at night they appear dark. Some fish appear flat. They look very wide when seen from the side, but very thin when seen from other angles. They are not easily seen by predators above or below.

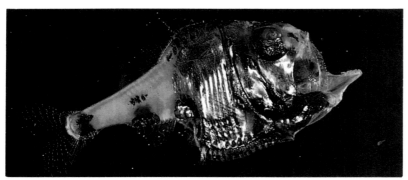

▲ The hatchetfish is well designed for hunting in the twilight. How many special features can you spot? Answers on page 152.

Finding food in the dark

Smell and taste are important senses at these depths. Deep-sea fish like the rat-tail fish often have long tails. These contain long lines of sensors, which detect vibrations in the water caused by the movement of other animals. The tripod fish (right) lives on the seabed. It has three extra-long fins which it uses as a tripod to prop itself up above the muddy waters so it can smell for prey in the clearer water above.

Food from above

All living creatures depend on plants for food. Either they eat plants, or they eat the animals that eat plants. The ocean is an unusual place because the plants can live only in a small part of it — near the surface, where there is enough light for photosynthesis. The animals in the rest of the ocean have to live either on other animals, or on the remains of dead organisms that sink down through the water.

Looking up

Predators try to see the silhouette (the dark outline) of their prey against the light coming from above. Many deep-sea fish have eyes that are directed upward.

The prey need to camouflage their bellies. Often they use light-producing organs. Rows of tiny pockets of bacteria produce light by chemical reactions. This makes the fish invisible against the light from above.

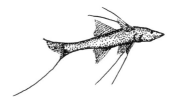

▲ Deep-sea anglerfish with luminous lure.

Animals that make light

In deeper water, where there is no light at all, many animals produce their own light. They do this to see their prey, or to deceive their predators. Anglerfish have luminous lures which they wiggle to attract other fish. When the fish comes within reach, the anglerfish opens its large mouth and sucks them in. Some tiny shrimps have spots of light on very long antennae, which make them seem much larger, so that predators will not attack them.

Some of the creatures in these dim waters move up to the surface to feed at night, when predators cannot see them. The lanternfish and the flashlight fish use light organs under their eyes like searchlights to find their prey.

◀ The deepest parts of the ocean have not been fully explored yet. Here, the environment has not changed for millions of years. Some of the animals found here cannot even be classified. They are not like any other living animals. This strange community of animals is found around hot springs, at great depths, near the Galápagos Islands.

Big mouths and stretchy stomachs

There is not so much food in the deep ocean, so the animals living there are few and far between. Many of the fish have very large mouths and elastic stomachs to take whatever prey they can catch. Some can even unhinge their jaws to catch prey larger than themselves. Their teeth curve backward, making sure that once caught, the prey does not escape.

Did you know?

There are more than 10,000 different species of copepod.

The giant squid is the largest invertebrate in the world, up to 47 ft long. It lives in the deepest parts of the ocean.

More about life in the oceans (pp. 116, 117); camouflage (pp. 52, 53); animal senses (pp. 50, 51)

Coral reefs

What are coral reefs?

Coral reefs are the most massive structures any living creatures have ever made. They are made up of the limestone skeletons of millions of reef animals, cemented together by sand. Living corals form only a thin layer on the surface of the reef. The dead coral rock underneath may be thousands of feet deep. Reef-building corals need light and warmth to grow, so living coral reefs are found only in shallow tropical seas.

Sand producers

Many fish, sea urchins, starfish and sea slugs feed on the corals. The parrotfish scrapes at the coral using its hard, beak-like teeth. As it feeds, bits of coral limestone fall off to form coral sand, which fills in cracks in the reef. Some sponges and clams bore into the rock for protection, producing more coral sand.

Corals are animals

Corals are animals. Each animal is cup-shaped and hollow and is called a polyp. It has a mouth at the top, surrounded by a ring of stinging tentacles which it uses to catch its prey. Many corals feed only at night. By day, the polyps shrink back into the cup for protection.

Corals are plants

The corals' tissues contain tiny plants called algae. For every square foot of coral surface there may be 139 billion algae below. The algae produce food and oxygen by photosynthesis. Some of the food and oxygen is passed to the coral. In turn, the coral protects the algae, and provides them with carbon dioxide. This extra food makes it possible for the coral to grow fast enough to produce the massive coral reefs. The algae need light for photosynthesis. This is why corals grow only in shallow water where a lot of light reaches them.

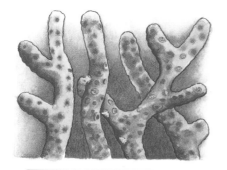

Hidden reef dwellers

Thousands of small animals live in the crevices in the reef. Crabs and shrimps scavenge for food among the corals. Butterflyfish use their long snouts to probe the crevices to get at them. Bristleworms and ribbon worms can wriggle into the thinnest cracks in the reef. Starfish are everywhere. They attack even the hard-shelled clams and scallops. Many reefs have large caves, and bigger predators lurk here — nurse sharks and octopuses, and the huge moray eel.

Some reef animals feed at night, and stay hidden during the day. Others have a wide range of camouflage. Some are colored to match their background, or are shaped to look like fronds of seaweed. Many have stripes which break up their outlines, and some even have false eyespots to direct the predators to less vulnerable parts of their bodies.

Corals are minerals

Corals produce a limestone skeleton around their bodies. They form colonies by budding off new polyps, which do not completely separate from each other. There are many different shapes of colony. Some are large and branching, some are delicate and like large leaves, while others form massive boulders. The beautiful sea fans are also corals.

A paradise for fish

All the thousands of small animals of the reef produce even smaller eggs and young. There is so much food on a reef that large fish shoals can be found everywhere, and bigger fish, such as reef sharks and barracuda, come to feed on them.

Corals are the oldest living animals on Earth. Some colonies may be several hundred years old, and contain over one million polyps.

▲ These beautiful soft corals do not form hard skeletons. Instead, they have jellylike coats with little spikes of limestone embedded in them for extra support.

Filter feeders

Many of the reef creatures filter food from the water. Sponges and sea squirts form colorful crusts on the coral rock, fan worms and barnacles sweep their bristly combs through the water, and giant clams draw currents of water through their shells. At night, feather stars climb to the top of the sea fans and sweep their arms through the water to filter out food particles.

Life on islands

Islands: large and small

There are many different kinds of islands. There are large ones and small ones. Some are covered with forests, and others are just bare rock.

Islands are formed in three main ways. Some islands appear when an underwater volcano erupts. The islands are formed from hot liquid rock, from the volcano, which solidifies in the sea. Many tropical islands are made of coral which builds up slowly in the sea. These islands take many thousands of years to form. An island is sometimes produced because the action of the sea begins to separate a small piece of land from the mainland.

Carried on the wind?

Many islands are a long way from land. Because of this, they are often difficult to reach. Even so, most islands have some plants and animals living on them. Scientists are interested in finding out how these plants and animals reached the islands.

Many microscopic plants and animals are carried by the wind. These tiny organisms are called aerial plankton. When these organisms are blown onto islands, some of them survive and begin to live and grow there. We say that they colonize the island.

▲ Insects, birds and bats are able to fly long distances. These animals can sometimes reach even the most distant islands. Many birds use islands as resting places on their long migration routes.

Seeds on the move

Birds and mammals often carry plant seeds, either clinging to their feathers or fur, or inside their digestive systems. In this way, many seeds reach islands. Later, they begin to grow into plants. The coconut is one of the best colonizers of islands. It floats inside its own husk which is resistant to sea water. Coconut palms are found on many tropical islands.

Adapting to needs

Many plants and animals become specially adapted to life on islands. A species of cormorant on the Galápagos Islands in the Pacific Ocean has very small wings. Over thousands of years its wings have gradually become smaller. Now they are too small for it to fly. Another bird, called a rail, lives on the island of Aldabra in the Indian Ocean. It, too, is flightless. Many islands have few large predators. Scientists think that some birds became flightless because there is no need to escape from predators.

Islands often have very large animals and plants living on them. The big island of Papua New Guinea is the home of giant birdwing butterflies. Giant tortoises and huge cacti are found on the Galápagos Islands, and on Aldabra Island. The world's biggest lizard lives on the island of Komodo in the East Indies.

Too heavy to fly
Insects called lacewings live on some of the Hawaiian islands. Their huge wings are too big and too heavy for flight. This stops them being blown away by the wind.

▲ Islands form safe places for arrivals to breed and bring up their young. Many islands have large colonies of birds, seals and even walruses. Turtles come ashore on some islands to lay their eggs in the sand.

Island stowaways
Large animals, such as reptiles and mammals, may float across the sea on rafts of vegetation. Sometimes these rafts reach islands by chance, and the stowaways come ashore, and survive. Large animals often carry smaller animals in their fur or feathers. In this way, many small animals, such as insects and spiders, are carried to islands hundreds of miles from land.

Did you know?

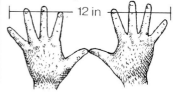

The world's biggest butterfly, the Queen Alexandra birdwing, lives on Papua New Guinea. It has a wingspan of almost a foot.

The coco-de-mer is the biggest plant seed. It grows on only one island in the Seychelles. It hasn't reached other islands, unlike its relative the common coconut.

The world's most remote, inhabited island is Tristan da Cunha. It is 1,700 mi from the coast of Africa.

More about aerial plankton (p.88); adapting to places (pp.20, 21); migration (pp.78, 79); size (p.27)

Life in caves

▲ Stalactites and stalagmites are columns of lime which are found in caves. They are formed by dripping water which contains dissolved calcium carbonate. Some of the drips stay on the roof of the cave where they form stalactites as the water evaporates. Other drips fall to the floor where they leave little deposits of lime. These form stalagmites. They "grow" very slowly, adding only a tenth of an inch each year.

A cave usually has three zones or regions. The part nearest the outside is called the light zone. Farther inside the cave it becomes dim. This is the twilight zone. The deepest part of the cave is called the dark zone. It is a very stable environment where temperature and humidity never change.

"Red-hot" tunnels

The lava tube (above) is in Hawaii. It was formed by volcanic action. When a volcano erupts it sends a stream of lava flowing down the side of the mountain. This is called the lava flow. The lava flow gradually cools down as it moves along. The outside cools first and soon becomes solid. The inside remains molten and continues flowing like a red-hot river. The solid lava on the outside insulates the hot "river" inside and stops it cooling. When the eruption stops, the flowing hot lava drains away leaving behind a long tunnel or lava tube. After the lava has cooled, some animals and plants are able to live there.

The world's biggest cave

Caves and underground tunnels are usually found in limestone rocks. Limestone is quite a soft rock, and it is easily eaten away by water containing dissolved carbon dioxide. As the water trickles through the limestone it begins to make small holes and tunnels. These slowly become deeper and bigger, and caves and caverns begin to form. This takes many thousands of years, and the caverns vary in size. The biggest "cave room" so far discovered is in Sarawak, in southeast Asia. It was found by an expedition in 1980. It is 2,300 feet long, and its average width is 984 feet. Its roof is more than 230 feet high, and the whole cavern is big enough to park about 20,000 cars.

Light to twilight

Caves are home for a number of different animals and some plants. It is not easy living in a cave. One of the main problems is finding enough food. This becomes more difficult the deeper you go. There is plenty of food in the light zone. Plants can grow in this zone because there is enough light, and large numbers of insects are found here. Earthworms and snails live in this zone and so do frogs and toads. Other animals visit the light zone from time to time. There is less food in the twilight zone and fewer animals live here. Bits of dead animals and plants are brought in by the underground streams or are washed in by rainwater. This provides food for shrimps and little animals called isopods, which are like woodlice. There are even small fish living in the underground streams.

Prisoners of the dark

Caves are very unusual habitats and only specially adapted animals can live in them. In fact, these animals cannot live anywhere else. They are "prisoners" of the dark. The animals include certain spiders, crickets, centipedes and fish. Even a crayfish, a kind of freshwater lobster, lives here. Animal colors don't seem to form in the dark. Most of the dark zone species are colorless. They also have poorly developed eyes, or no eyes at all. Eyes are not much use in complete darkness. Instead, these animals have long antennae and use their sense of touch to feel their way about. Some of the fish can detect tiny vibrations in the water.

The "bat" bird

The oil bird lives in caves in parts of South America. Like bats, it uses the caves only for sleeping in during the day. Oil birds roost in large numbers. The oil bird has very good eyesight, but its eyes are not much use in complete darkness. Even so, it can fly around quite easily in the dark because it navigates in the same way as bats. This is why it is sometimes called the "bat" bird. At night the birds leave their caves to feed on oil palm fruits.

▲ Bats are part-time lodgers in the dark zone. They use caves to sleep in during the day. The bats hang upside-down from the roof in large numbers. Their droppings collect on the cave floor and provide food for other animals.

Did you know?

The Proteus salamander lives in caves in Yugoslavia. At birth it has eyes and its body is colored. However, it loses its eyes and its color as it gets older. It never really becomes adult, and it reproduces in the tadpole stage. It can go for several years without food.

More about how bats navigate [p.80]; animal senses [pp.50, 51]; hunting in the dark [pp.126, 127]

Night life

Life after dark

Many animals are active at night. We say they are nocturnal. Some plants even open their flowers at night and close them during the day. Animals are nocturnal for different reasons. Some only come out at night to avoid predators. Others hunt their prey at night because this is when it is moving about. Some animals are out at night because it is either safer or more comfortable. Spiders spin their webs at night, and many butterflies and moths emerge from their pupal cases under the safe cover of darkness. Earthworms come up to the surface of the soil at night to feed. At this time, the humidity is higher and there is less chance of the earthworm's skin drying out. Desert animals usually become active after dark when the air is cooler.

Eye-spy
Most of the world's 130 species of owls are nocturnal. They have huge, forward-facing eyes used for hunting their prey in dim light. An owl's eyes are so big that they occupy most of the space in each eye socket in the skull. Because of this, there is no room for any eye muscles. An owl doesn't move its eyes. Instead it moves its head. The flexible neck allows the owl to turn its head either way to face backward without moving its body. Other nocturnal animals such as tarsiers can also do this.

▲ One of the main problems facing nocturnal animals is finding their way. They solve this problem in different ways. Some, such as owls, bush babies and geckos, have large, sensitive eyes designed for seeing in dim light. Night eyes can open very wide to let in the maximum amount of light. They can also become very narrow in daylight.

▲ Nocturnal snakes such as pit vipers have pits on the head which sense the heat given off by the bodies of other animals. They use these organs to find their warm-blooded prey in the dark. Heat detectors only work because the prey's body is warmer than the snake's. Snakes can also detect vibrations on the ground through their chins.

Big-ears
Nocturnal animals often have large, sensitive ears which they use for detecting prey and predators in the dark. Bats find their way and locate food by sonar. Their high-pitched cries bounce back off objects in their flight path. These echoes are picked up by their large, sensitive ears. The frog-eating bat from Central and South America finds its prey by picking up the frogs' night calls with its sensitive ears.

◀ This Australian frogmouth matches the bark on which it sleeps during the day. Many nocturnal animals are camouflaged to help avoid day predators. The South American potoo blends with the tree on which it roosts. It looks like a large, dead branch. Insects such as mantids and moths remain motionless during the day. Many tree frogs sleep with their legs drawn up and their eyes closed. This makes them more difficult to see.

Night "talk"

Night life has developed various ways of communicating in the dark. Because vision is limited, many animals "keep in touch" by sound, or by light signals. Insects such as crickets become very active at night. The males "sing" for long periods by rubbing one wing over the other. Temperature affects the rate of chirruping. The speed of a cricket's song increases on warm evenings and decreases when it is cooler. Frogs also join in the night chorus, especially in the tropics. They often have large throat sacs which produce and amplify the noise they make.

Lighting-up time

Fireflies and glowworms are light-producing beetles. In both cases the light is for sexual attraction. Only female glowworms light up to attract the males. The male fireflies give off short bursts of light to attract females. Light-producing organs usually have a reflecting layer and a series of lenses to increase the strength of the signals.

South American tree frog

Many plants have flowers which open at night. These flowers are usually pale so they are easily seen in the dim light. This makes it easier for animals like the honey possum to find the plant and help pollinate the flowers. Night flowers also have a strong scent to help attract moths and bats. Flowers pollinated by bats usually hang free of the leaves so bats can easily find them by sonar.

Did you know?

The click beetle from the West Indies has an orange light that it turns on when taking off and landing.

The heat pits of a rattlesnake can detect a small mammal only 11.7 in away.

In the Second World War, soldiers in the jungle of Malaysia carried branches with light-producing fungi on them. They acted like torches in the dark.

More about animal senses [pp.50, 51]; communication [pp.72, 73]; attracting a mate [pp.64, 65]

Wildlife goes to town

City dwellers

A large town, or city, with its many houses and factories does not seem a likely place to find wildlife. For years scientists paid little attention to the animals and plants living in our towns. Instead, they studied plants and animals in the countryside, and those in remote habitats. Recently, scientists have become more interested in town habitats, and in the plants and animals living in them.

▲ Some mammals have become very successful town invaders. Foxes are now quite common in many towns. They bring up their cubs in towns. Hedgehogs are also common town animals. They live in gardens where they are able to find plenty of food. In American cities, some raccoons raid dustbins at night. And some settlements in northern Canada are even visited by polar bears.

Where do the invaders come from?

All animals and plants living in towns have come from the surrounding countryside. Some animals enter the town and find a habitat like the one they left in the country. Many of our garden birds do this. Others find a very different habitat, and adapt to the new conditions.

The good life

Wildlife may find that living in a town has certain advantages over living in the country. People and buildings produce a lot of heat, which makes the town a warm place in which to live. The air temperature of a town in winter is higher than that of the surrounding countryside. Many animals are attracted by this warmth. The buildings of a town also give shelter from the wind. Some buildings serve as safe places where animals can breed. Food is in plentiful supply in the town's dustbins and rubbish tips. Many wild plants also find a home in the gardens of the town.

The artful alligator

In parts of Miami, in Florida, alligators have invaded private swimming pools, and lakes on golf courses. They have started to do this because their natural habitats are being destroyed by human activity.

Water, water everywhere

Big towns and cities have plenty of watery habitats. There are lakes, rivers and reservoirs. In addition, many gardens have artificial ponds. All of these are places where aquatic plants and animals can live. The large lakes and reservoirs attract a lot of birds. The small garden ponds are good places for frogs, toads and newts to breed. Garden ponds are often stocked with fish of various kinds. Birds such as herons are attracted to the ponds to feed on the fish.

The big cleanup

In recent years pollution of many town rivers has lessened or stopped altogether. Laws have been passed which prevent factories from emptying poisonous chemicals into town waterways. The rivers are now much cleaner, and new species of fish are beginning to invade them. In recent years, more than one hundred species of fish have come back to the River Thames in London.

▲ Birds are well adapted to town life. Some birds are more common in towns and cities than in rural areas. Starlings and pigeons live in very large numbers in many cities. Sometimes their numbers become so great that they begin to be a nuisance. They make a lot of noise, and they damage buildings with their droppings.

Mice in cold store

House mice have been reported living in cold meat stores, where the temperature never rises above the freezing point. They have adapted to these cold conditions and even breed in them. These freezerstore mice have thicker fur, and are bigger than their relatives who live in ordinary conditions. Can you explain why? Answer on page 152.

The rat invasion

Rats and mice are the most common town mammals. Some towns may have more rats than humans living in them. The rats cause damage and may spread disease.

Did you know?

In Venice, there are so many pigeons that they have been given birth control pills to keep their numbers down.

The chimney swiftlet once nested in hollow trees in North America. Now it prefers hollow ventilator shafts and disused chimneys in towns and cities.

Ospreys nest in sight and sound of the space shuttle launch pad at the Kennedy Space Centre in Florida in the United States.

More about adapting to places (pp.20, 21); pond life (pp.110, 111); size (p.27); life in cold places (p.100)

Tropical rain forests

Hot, wet and never changing

Tropical rain forests grow on lowlands and foothills in regions close to the Equator. They are found in equatorial Africa and South America, and there are smaller forests scattered throughout southeast Asia and Australasia. In these regions there are no seasons, and day and night are equal in length all year round. The air temperature remains between 68 and 86°F throughout the year, and the rainfall is always greater than 80 inches per year. These hot, wet conditions are just right for rapid plant growth. Because there are no seasons, conditions in these regions never vary. This makes tropical rain forests very stable places for animals and plants to live in. The forests have been like they are today for millions of years. They are the richest habitat on Earth for wildlife and they contain an enormous variety of animals and plants. The picture shows the main layers which make up a tropical rain forest.

A few trees grow so tall that they stand well above the canopy layer. The atmosphere here is fresh and windy, and **emergent trees** use the wind to distribute their seeds.

The **canopy** is a dense and continuous layer of green leaves about 25 feet deep. Many animals live here, feeding on the rich supply of leaves and fruit.

The **understory** receives very little light. It is the "highway" along which many animals travel when moving through the forest from floor to canopy.

The **soil** in tropical rain forests is not very deep. The giant emergent trees put out shallow roots to catch the nutrients before the heavy rains wash them away. They also produce huge buttress roots which help to support their great height.

The **forest floor** is semidark and few plants grow here. It is covered with dead leaves. The humid conditions encourage decay. Fungi, ants and termites live here in large numbers.

54 yd

emergents layer

41 yd

canopy layer

27 yd

understorey layer

14 yd

ground

buttress root

130

▲ Each leaf in the canopy is positioned to catch the maximum amount of sunlight. Many leaves can even change their position during the day as they follow the Sun's path across the sky. The leaves have 'drip tips' which act like the spout of a jug and help the heavy rain-water to drain away quickly. They also have a waxy, protective covering which prevents the spores of small plants, such as moss and algae, from getting a hold and germinating on their surface.

Life in the forest canopy

Most of the animal life in a tropical rain forest lives in the canopy layer. It is only recently that scientists have been able to climb into this part of the forest habitat, 80 feet above the ground, and study the wildlife living there. The canopy forms the rain forest ceiling. The dense foliage encloses a warm, humid atmosphere which supports a variety of animals. There are browsers, such as sloths and monkeys, feeding on leaves and fruits. Hunters like the monkey-eating eagle feed on the browsers. There are also scavengers, and even some animals which steal food from others. When the animals of the "day shift" have finished, nocturnal animals take over. Pottos, bush babies and lorises climb through the canopy. They have huge eyes to find fruits and insects in the dark. Large fruit-eating bats arrive and flap among the leaves, looking for ripe fruits on which to feed.

Climbing and hanging on

Life in the canopy requires special methods of locomotion. Birds can flap, jump or fly from one branch to another. Some clamber about the branches in search of food. Certain mammals have developed a tail for gripping, and extra-large claws for hanging on to branches. It is also important to be able to judge distances very accurately. Because of this, many mammals have developed binocular vision for finding their way among the branches.

The living coat-hanger

The two-toed sloth lives in the rain forests of South America. It is found in the canopy where it does everything upside-down. It feeds on fruit and leaves, and even sleeps, while hanging by its feet from the branches. The sloth is completely adapted to an upside-down life in the canopy. It hangs by stiff, rodlike legs and moves very slowly, gripping the branches with its hooklike claws. Even its fur grows in the opposite direction to that of other mammals so the rain can drip off easily. Green algae grow on its fur and these camouflage the sloth from its predators. It can move its head through nearly 360°, allowing it to see all round without moving its body.

Did you know?

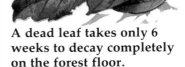

The South American rain forest covers an area of 965,000 mi².

A dead leaf takes only 6 weeks to decay completely on the forest floor.

The world's tropical rain forests are being damaged or destroyed at the rate of 99 acres per minute.

More about rain forest animals [pp.132, 133]; forests [pp.10, 11, 106, 107]; binocular vision [p.50]

Animals of the tropical rain forest

FOREST RANGER'S NOTEBOOK

The tropical rain forest is one of the richest habitats in the world. It contains a wide variety of animals and plants. Although some large animals live on the forest floor, most animals live in the canopy, 80 feet above the ground.

1 **Toucan** Lives in rain forest of South America. Canopy dweller. Feeds on fruit, insects and even eggs.

2 **Howler monkey** Lives in groups in South American rain forest. Very noisy. Calls heard over a mile away.

3 **Great blue turaco** Found in African rain forest. Noisy bird. Feeds on fruit and small grubs. Often travels in groups.

4 **African gray parrot** Lives in flocks in canopy of African rain forest. Feeds on fruit and seeds. Powerful beak for breaking seeds.

5 **Aye-aye** Lives in forests of eastern Madagascar. Nocturnal. Feeds on fruit and insect larvae.

6 **Malay fruit bat** From rain forest of southeast Asia. Largest of bats. Feeds mainly on bananas and figs.

7 **Slender loris** Lives in forests of southern India and Sri Lanka. Nocturnal. Feeds on insects and fruit.

8 **Lar gibbon** Found in canopy of rain forest of southeast Asia. Swings through canopy using its long arms. Feeds on fruit and some insects.

9 **Flying frog** Found in rain forest in Borneo. Has enlarged webs between toes. Can glide downwards using webs as parachutes.

10 **Orangutan** Lives in Borneo and Sumatra. Feeds on fruit, leaves and insects. Lives alone or in family groups. Expert climber.

11 **Colugo** Found in forests of southeast Asia. Glides on large body membranes. Nocturnal. Lives on leaves, fruits and flowers.

12 **Great hornbill** Canopy dweller. Large beak, long eyelashes. Feeds in flocks on fruits and nuts. Southeast Asia.

13 **Six-wired bird of paradise** Lives in rain forest of New Guinea. Male does special dance for female on forest floor.

14 **Okapi** Rare animal from African rain forest. Not discovered until 1900. Feeds mainly on leaves.

15 **Royal python** Lives in equatorial West Africa. Maximum length 5 feet. Curls up into a ball when attacked.

16 **Lowland gorilla** Lives in family groups in rain forest of West Africa. Adults too heavy to climb trees.

17 **Common iguana** Lives in forests of South America in trees along river banks. Feeds on fruits. Good swimmer.

18 **Tamandua** Found in rain forest of South America. Nocturnal. Uses tail to grip branches. Feeds on ants and termites.

19 **Spider monkey** Found in tropical American forests. Swings through canopy on long arms. Tail used to grip branches. Feeds on fruit.

20 **Kinkajou** Found in forests of Central and South America. Nocturnal. Mixed diet including plants, insects, mammals and birds.

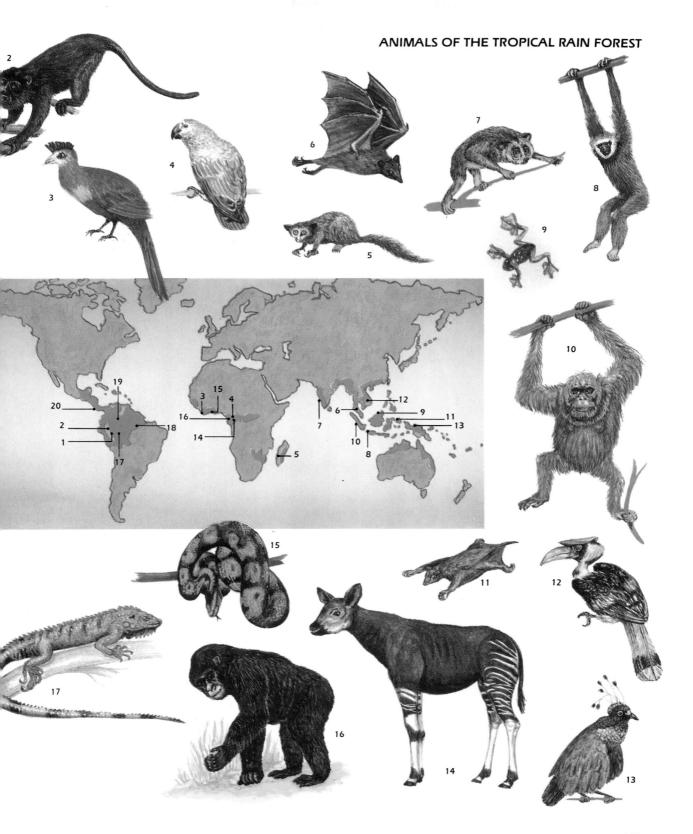

Endangered species

CONSERVATIONIST'S NOTEBOOK

Many animals are in danger of becoming extinct. This is because their habitats are gradually being destroyed by humans. The animals shown here are all endangered species.

1 **Californian condor** Found in mountains northwest of Los Angeles. Total population about 6.

2 **Polar bear** Found in Arctic regions around North Pole. Total population about 12,000.

3 **Spanish lynx** Found in one small area of southern Spain. Population perhaps a few hundred.

4 **Walrus** Lives in Arctic region. Hunted by Eskimos. Atlantic race disappearing. Population 25,000.

5 **Giant panda** Found in mountains and forests of Tibet and southwest China. Probably less than 1,000.

6 **Caspian tiger** Found on border between Russia and Afghanistan. Farming has destroyed habitat.

7 **Monkey-eating eagle** Lives in Philippine Islands. Becoming very rare. Total population about 40.

8 **Orangutan** Found only in parts of Borneo and Sumatra. Total population about 5,000.

9 **Przewalski's horse** Found in southwest Mongolia. Very rare. Probably more in captivity than wild.

10 **Komodo dragon** World's largest lizard. Lives on 4 small islands near Indonesia. Only a few hundred exist.

11 **Tuatara** Rare burrowing lizard. Lives on islands near New Zealand. Perhaps 20,000 survive.

12 **Kakapo** A large flightless parrot from New Zealand. Nocturnal. Less than 100 survive.

13 **Tasmanian "wolf"** Thought still to exist in Tasmania. Tracks found and sightings made.

14 **Koala** Found in eastern Australia. Recently, population reduced because of virus disease.

15 **Indris** Largest of lemurs. Found in northeast Madagascar. Lives in forest habitat. Vegetarian.

16 **Cheetah** The African cheetah is becoming rarer. Total population probably less than 25,000.

17 **Great Indian rhinoceros** Found in northwest India and Nepal. Population about 500.

18 **Mountain gorilla** Found near Virunga volcanoes in equatorial Africa. Population about 400.

19 **Leatherback turtle** Largest of turtles. Once common in tropical seas. Breeding sites threatened.

20 **Blue whale** Biggest animal that has ever lived. Suffered from overhunting. Population less than 2,000.

21 **Green turtle** Used to be common in most tropical seas. Hunted for food. Eggs taken from breeding grounds.

22 **Giant armadillo** Biggest of the armadillos. Lives in South America. Threatened by increased farming activities.

23 **Flightless cormorant** Found on only one of the Galápagos Islands. Eggs collected. Also hunted for food.

24 **West Indian manatee** Found near coasts of Caribbean Sea and northeast South America.

25 **Southern sea otter** Thought to be extinct but rediscovered in 1938. Population now about 2,000.

Conservation — saving wildlife

▲ Great areas of forest were cut down when the Trans-Amazon Highway was built. This also caused large numbers of animals to disappear.

As dead as a dodo

You have probably heard the phrase "as dead as a dodo." The dodo was a species of bird. The last dodo was killed on the island of Mauritius more than three hundred years ago. Since then, many other species of animals and plants have disappeared forever. When this happens, we say the species has become extinct. Animals and plants become extinct naturally. This is part of the process of evolution. However, in recent times, these disappearances have increased, and humans have been responsible for most of them.

Farms, industry, hunting and poaching

Many of the world's natural habitats are gradually being destroyed. More land is being claimed for farming. Areas of forest have been cleared and land drained to make room for factories. Land has even been flooded to make huge artificial lakes to provide hydroelectric power. Forests are cut down for timber and to make room for roads and highways. Drilling for oil has destroyed habitats on land and at sea. Other threats to wildlife are hunting and poaching. More than one million whales have been killed in the last eighty years by hunting. Poachers in Africa have reduced elephant and rhinoceros populations to near extinction.

Passenger pigeons

About one hundred years ago, the passenger pigeon was the most common bird in the world. One flock on migration was estimated to contain more than 2 billion birds. The pigeons were tasty and were hunted for food. In 1869, 7½ million birds were caught at one spot. In 1879, 1 billion birds were captured in the state of Michigan in the United States. The last passenger pigeon died in Cincinnati Zoo in 1914, and so the passenger pigeon became extinct.

Every year, millions of animals and plants are shipped around the world. Many end up as pets and house plants. Other animals are killed and their skins used for furs and leather. Elephant tusks are made into trinkets and the wood of many rain forest trees is used for building.

▶ Operation Tiger started in 1972. Its main aim has been to find out how to protect the Indian tiger. Over the years, tigers have been moved from places where they are in danger to game reserves where they are protected. The project has been very successful and the Indian tiger has increased in numbers once again.

DON'T SQUASH ME!

Safe places to live

Many governments in different parts of the world have put aside areas of land for the preservation of wildlife. These areas vary in size. Some are quite small, but others cover many thousands of square miles. The Wood Buffalo National Park in Canada covers 17,560 square miles. The Etosha Reserve in Namibia, in Africa, is the world's biggest reserve, covering 38,427 square miles. North America and Canada have more than one hundred and sixty game parks and nature reserves. In these reserves wildlife can live without being disturbed, and hunting is banned. Park rangers and wardens stop poachers from killing the animals. Even so, many of the world's animals and plants are still in danger of dying out. Because of this, special projects have been set up by the World Wildlife Fund and other organizations to save those animals most in danger.

Panda problems

The giant panda is the symbol of the World Wildlife Fund. There are probably less than 1,000 still alive in the mountains of central China. In 1975, many pandas starved because the bamboo plants on which they feed flowered and died. This put the giant panda in great danger. In 1980, the Chinese government began work with the World Wildlife Fund to study the giant panda in its natural habitat. Animals were fitted with radio transmitters so their movements could be followed. Scientists are finding out more about the pandas' way of life.

HELP SAVE US

WE ARE MOUNTAIN GORILLAS THREATENED WITH EXTINCTION BY POACHING AND FOREST DESTRUCTION WE LIVE ONLY IN RWANDA, ZAIRE AND UGANDA
THERE ARE FEWER THAN 400 OF US LEFT
HELP
THE FAUNA AND FLORA PRESERVATION SOCIETY
RAISE FUNDS FOR THE
MOUNTAIN GORILLA PROJECT
Information/donations: Fauna and Flora Preservation Society, c/o ZSL, Regent's Park, London NW1 4RY

Only about 400 mountain gorillas survive in the rain forests of Central Africa. The Mountain Gorilla Project was set up in 1978 to protect these animals and help them survive. People living in the area have gradually cut down the gorillas' forest home to make more farming land. The Rwanda Government works closely with the Project to educate local people about conservation of the forests. A team with a special van goes to schools and villages to show films and give talks about the importance of the rain forest and its gorillas. School trips have been arranged to see the gorillas in their natural habitat, and wildlife clubs have been set up in local schools.

Zoo news

Old and new

The first zoos were very different places from the zoos of today. In the beginning, people did not care about the animals, how they were kept or the cages they lived in. The animals' comfort was not very important to the zoo owners, or the visitors. If any of the animals died, they could easily be replaced by catching more from the wild.

Today the situation is different. Many animals are rare in their natural habitats, and some have disappeared altogether. Several governments have banned the export of wild animals. For these reasons it is much more difficult for zoos to obtain new specimens. If a particular animal dies, the zoo cannot replace it easily unless it can breed its own specimens, or buy from another zoo.

◀ This is a picture of the Tower of London's menagerie in about 1820. The animals were kept in "dens" under the arches. Compare this zoo with the modern one in the photograph below. If you were an animal, which zoo would you prefer to live in?

Moonlight world

Until recently, nocturnal animals like this Australian spiny anteater were never seen in zoos. During the day, they slept in a corner of their cage and only became active after all the visitors had gone home. New ideas about keeping animals changed all this. "Moonlight" houses are now found in most big zoos. Here day and night are reversed. The animals inside live in a moonlit world when there is daylight outside. This allows visitors to watch nocturnal animals moving about and feeding. At night, the lights are turned on fully so the animals think it is daylight.

◀ In a modern zoo, animals are kept in well designed enclosures in surroundings similar to their natural habitats. Animals were often kept by themselves; but now many live in small groups. They are much happier like this. They are healthier and live longer. They also breed more easily.

Breeding rare animals

ne-ne

As animals in the wild become rarer, zoos are becoming more important as places for breeding endangered species. For some very rare animals zoos may be their only hope of survival. There are probably more Przewalski's horses in captivity than in the wild. Capturing the few remaining Californian condors and keeping them in zoos may save them from extinction. Zoos set up breeding programs for rare species. The ne-ne, or Hawaiian goose, has been successfully bred in captivity and even returned to the wild. In the future, zoos will probably become more involved in this kind of work. Perhaps history will be reversed. In the past zoos have taken animals from the wild. In the future they may give some animals back.

▲ Like many zoos, Paignton Zoo in Devon has an educational center where young people can learn more about the animals in the zoo. Programs are organized for school parties, and there are young naturalist clubs for children to join. Workers at the center give talks and show films. There are many other activities to help young visitors get more from a day at the zoo.

Zoo scientists

Zoos are not just places where animals are kept and bred. Many big zoos have large laboratories where scientists work to find out more about things like animal diseases and health. They also investigate animal diets in order to discover the best ways of feeding zoo animals. Much of what they find out helps other scientists working with animals in the wild.

Feeding time

Usually animals in zoos are not given the same food they would get in the wild. They are fed on carefully worked-out diets. Here are some daily menus.

ELEPHANT'S MENU
99 lb HAY
4 lb MEAL
26 lb VEGETABLES
4 lb BREAD
9 oz SALT

GIRAFFE'S MENU
4 lb OATS
7 lb VEGETABLE
26 lb HAY
5 oz SALT

BROWN BEAR'S MENU x 2 A DAY
15 lb MIXED RICE
MEAT + BREAD
4 lb VEGETABLES
APPLES

TOUCAN'S MENU
5 oz ALMONDS
HAZELNUTS
CHICKPEAS
4 oz BREAD + MILK
9 oz BANANAS

More about endangered species (pp.134, 135); nocturnal animals (pp.126, 127); mountain gorillas (p.19)

Looking for clues

Nature is all around you

Now that you have read this book, you will probably want to go out and look at nature for yourself. But at first you may be disappointed because you may not see any animals. Yet the signs of animal activity are all around you, if you know how to look for them. You have to look for clues. Once you start following the signs, you will probably start to see the animals themselves.

▲ Animal burrow in heather.

Be quiet

All animals have enemies. As humans are bigger than most animals, it is not surprising that animals hide away when they hear us coming. So try to be quiet. You will see more if you stop and wait than if you move about.

Listen!

Stop and listen. You can usually hear birds long before you see them. Every bird species has its own distinctive song. Birds sing most in the early morning, and in the late afternoon and evening. Listen for the alarm calls of birds that hear you coming.

Dropping clues

Animals have their own characteristic droppings. The droppings of a dog or fox are very different in size and shape from these gazelle droppings.

Collectable clues

Thorny thickets and barbed-wire fences are good places to look for other clues — hairs of badgers, foxes, deer and other animals. Feathers are easy to find and collect. Do you often find the same hairs or feathers in the same place? There may be a nest nearby, or you may be standing on a path used regularly by animals.

Nests and holes

The burrows of mice and foxes are easy to see. The size of the burrow will give you a clue as to who made it. Look for signs that it is still occupied — remains of food nearby, droppings, hairs. Use your nose — many burrows have a very distinctive smell. You may find well-worn tracks leading away from the burrow. Badgers follow regular paths when they go out foraging, and so do mice and rats.

Bird nests are also quite easy to find. Note the size of the nest, and what it is made of — the outer structure, and the soft inner lining. Are there any features to give you clues?

Look on the ground

Footprints are more easily found when the ground is soft, or if it is covered in snow. Good places to look are in the mud beside streams and ponds, in sandy areas, and areas where the soil is not covered by vegetation.

Note the size of the tracks, how many "toeprints" they have, and how far apart the footprints are. Are the tracks fresh? Can you follow the trail and get close to the animal that made it? Look at the detail of the tracks. Has the animal got long claws? Are its feet webbed? See if you can work out which tracks belong to the forefeet and which belong to the hind feet. Are there any other clues — feathers, hairs, droppings? Have its wings or tail left their tracks too?

Untidy eaters

Many animals leave the remains of their food on the ground. Different animals feed in different ways. Nutshells may be split in half by squirrels and hawfinches. Voles gnaw a round hole in nutshells. Dormice and voles leave a smooth edge to the hole, while wood mice leave tooth marks. Nuthatches and woodpeckers often wedge nuts in cracks in tree bark, either to hold them firm while they hammer them open, or to store them for later.

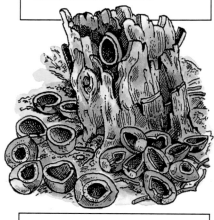

▲ Rabbits and hares put both their forefeet down at the same time. Deer put their feet down one after the other.

▲ Running and walking produce quite different patterns of tracks.

Beach combing

If you are at the seaside, the shore is a wonderful place for clues, not only to the life of the beach, but to life in the ocean beyond. Shells, eggs, skeletons, washed up jellyfish — all collect along the shore. The most exciting finds are often made after there has been a storm at sea.

◀ Looking at the ground can tell you what kind of plants there are above you. In tropical forests many plants grow on the branches of trees hundreds of metres up. Vines trail up other trees, but do not flower until they reach the top. But their flowers, fruits and leaves fall to the ground below, where you can see them.

More about animal homes (pp.60, 61); the beach (p.113)

Improving your nature photography

It's the ideas that count

There are so many different cameras that it is impossible in such a short space to tell you how to take good nature pictures with your particular camera. There are many good books on photography, including some which specialize in nature photography.

On this page, you will find ideas for making your pictures more interesting and fun. Even with a very basic camera, it is possible to take good nature photographs, good enough to publish. Your ideas can compensate for the limitations of your camera.

Lighting

You can get many interesting effects by varying the lighting. Instead of taking photographs with the Sun behind you, try lighting your subject from the side. Shadows can pick up the texture of skin or petals, and lots of small details which the rather flat lighting of bright sunlight does not show.

▲ Silhouettes can be impressive, and work well in black-and-white as well as in colour.

▲ Furry animals look quite dramatic when the light is coming from behind. The hairs show up as a bright line around the animal. Hairy plants produce the same effect.

If you have a single lens reflex camera, try underexposing your picture. You can turn day into night by doing this. Sometimes it produces very dramatic effects. It can show up striking cloud groups, or turn the Sun into a starry form with shafts of light coming from it. Underexposing is especially exciting in snowy conditions.

Coming in close

Even if you do not have the special equipment for taking really close-up pictures, you can get some interesting pictures by coming closer than usual to your subject. Think about photographing special features — feet, ears, eyes, noses, coats. A close-up of a zebra coat can be quite a puzzler for your friends!

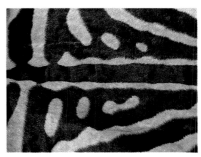

▲ Close-up of a zebra coat.

▼ Bold shapes and an unusual angle combine to make an interesting picture.

▲ Lichen patterns on tree bark.

Shapes and patterns

Look for interesting patterns in nature, such as arrangements of leaves, patterns on tree bark, the shapes of trees and branches in winter. Try and find bold shapes to frame your pictures. Overhanging branches, twisting tree trunks, and grasses in the foreground will all help to add interest and a sense of depth.

Practice can be fun

You can get a lot of practice of photographing nature by visiting local parks, botanic gardens, zoos and wildlife parks. These are good places to try comparing long distance and close-up pictures, and to sort out any problems. Keep going back to photograph the same animals again and again until you get your picture just right. Butterfly houses are marvellous places to try out close-up photography.

Looking through the lens

A final word of warning. It is difficult to take good pictures if you are trying to do other things at the same time. It is no good going out and trying to look at birds as well as taking photographs. To get good pictures you need to be thinking photography. You have to be looking at the world around you as if you are seeing it through the lens of a camera.

Watch your background

Watch your background. Walk around your subject to find the best angle from which to photograph it. Brightly lit flowers or animals look even better photographed against a dark background. Dark subjects may look better against a light background. If the background is cluttered, it may take away interest from your main subject. On the other hand, if the background is important, use a wide-angle lens to give you greater depth of focus — this means that more of the background is in focus.

▲ A wide-angle lens can really show the subject in its natural habitat.

143

Natural history art

Drawing animals and plants

You can learn a great deal about animals and plants by drawing them. To draw well, you must watch nature closely. It is this close study which leads to a greater understanding of the plants and animals that share the world with us. Here we show you how you can start drawing and painting living things.

Working against the clock

Time is very important. Animals move quickly. They seldom stand still, or "pose" for you. So you must learn to put the maximum amount of information on paper in the shortest possible time. Drawing from life is the best way for you to become familiar with an animal because you can study its behavior and movement. It is a good idea to draw a pet, or a farm animal, or even take a trip to the zoo to practice drawing quickly. When drawing a plant you will have more time to think about detail and use of color.

Composing the drawing

Sketching your subject as a series of shapes in the form of circles, cylinders and eggs will help you to draw its outline. Compare the height with the width, and note the distances from head to tail and from head to feet. You can make your observations more accurate by using the pencil as a measuring device.

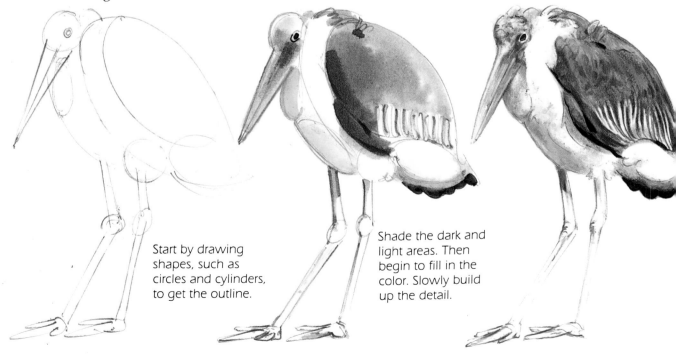

Start by drawing shapes, such as circles and cylinders, to get the outline.

Shade the dark and light areas. Then begin to fill in the color. Slowly build up the detail.

Black and white

Half-closing your eyes can help you to pick out which areas are dark and which are light. You can draw different depths of light and shade by using different pencils. A very soft 4B used on its side gives the rough texture of the coat of a squirrel or fieldmouse. A harder 2H pencil will give the smoothness of a beak, hoof or horn.

Use the eraser as little as possible, as it damages the surface of the paper. Mistakes can be covered by new lines or areas of tone.

Fur, hair or bushy tails are not easy to draw! This squirrel has been sketched using a very soft pencil to give the rough texture of its coat and tail.

Color

Having drawn the outline and general shape, choose the basic overall color. Cover the shape with a light coat of paint, except in the lightest parts. Using your crayons and pastels, build up the darker regions. Do not fill in the detail until the pattern of tones is completed. If a color is too strong it can be diluted with a little water.

Color can be used to fill in the details of your subject.

Identikeys

Putting a name to it

When a scientist is talking about an animal or plant, scientists from other countries often like to know which particular species is being described. But scientists from different countries speak different languages. So how do they communicate? When they give a new animal or plant a name they use an international language. The name they give to each species is a special one. It is not an ordinary name like "dog" or "cat." It is a special name, called the scientific name, and it is made up of two parts. The first is the genus part of the name. The second is the specific part of the name. The two parts together make up the species, or scientific, name. Let us look at a famous animal to see how this works.

New finds

New animals and plants are being discovered all the time. Each new species has to be classified and given a scientific name. Most of the world's biggest animals have probably been discovered, but there are still some surprises. In 1979 a new species of lizard was found on the Fiji Islands in the Pacific Ocean. Until 1979, nobody knew it existed.

tiger

leopard

jaguar

A lion is a member of the cat family. It is one of the big cats, which include tigers, leopards and jaguars. The big cats belong to the genus *Panthera*. This word forms the first part of a big cat's scientific name. The second part is the specific word which separates one big cat from another. The lion's second name is *leo*. So its full scientific name is *Panthera leo*.

Here are three more big cats. Can you guess which scientific name belongs to which animal? You may have to use your imagination. Look for similar words.
A *Panthera pardus*
B *Panthera onca*
C *Panthera tigris*
Answers on page 152.

Imagine that you have just returned from an expedition to Africa. While you were there you discovered a new species of big cat. The cat is remarkable because it lives completely in water. Your expedition leader has asked you to give your new species a scientific name. Here's a clue. Think of another word which means something to do with water. Answer on page 152.

146

How to identify animals and plants

When you want to identify an animal or plant you have to look for clues. There are lots of ways to begin. You can start by looking at pictures in a book. Sometimes a picture matches the plant or animal you are interested in and this helps you identify it. Sometimes you have to read a description, but you still have to look for clues. You may learn about the habitat in which a particular organism lives. Is it the same kind of habitat as the one where you found your animal or plant? You will probably collect a lot of evidence before you can be sure of your identification.

A B

C D

E F

Using a key is like a treasure trail. An answer to one clue leads you to the next clue and soon.

Identification keys

You can identify an animal or plant even if you haven't seen it before. This is something scientists often have to do. To help them they use a "key." Just as a key opens a locked door, a scientist's key solves the mystery of an unidentified animal or plant.

Remember Look carefully at each penguin in turn. Start at the beginning. Think about each pair of clues. Then decide which step to go on to next. Good luck!

Using a key

Jim is an Antarctic explorer. He is interested in penguins, but knows nothing about them. They all look the same to Jim. Maggie is an explorer friend who knows a lot about penguins. She decides to help Jim by making a key so he, too, can identify penguins. Here are 6 penguins and part of Maggie's key. Can you use the key to identify each penguin? Answers on page 152.

MAGGIE'S KEY

1 Penguin has long, thin beak — go to step **2**
Penguin has short, thick beak — go to step **3**

2 Penguin has D-shaped patch on head — Emperor
Penguin has comma-shaped patch on head — King

3 Penguin has completely black head — Adelie
Penguin has light stripes or patches on head — go to step **4**

4 Penguin has tuft of feathers on head — Rockhopper
Penguin has no tuft of feathers on head — go to step **5**

5 Penguin has thin line across face — Chinstrap
Penguin has light patch above eye — Gentoo

Wildlife quiz

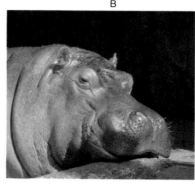

Try these knotty problems

Heads you win

Here are the heads of three different animals. Can you name them?

A
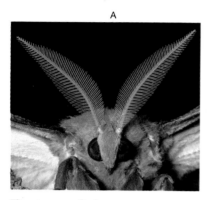

B

C

True or false

1 Birds have teeth.
2 Armadillos live in Africa.
3 Camels drink their own nose drippings.
4 Octopuses are mollusks.
5 Some insects can produce light.
6 Leopards live in South America.
7 There is a bat that can catch fish.
8 Sea horses are mammals.
9 All birds can fly.
10 Corals are plants.
11 Some insects and spiders look like bird droppings.
12 Some coral fish can live among the tentacles of sea anemones.

Food

1 What does the blue whale feed on?
2 Name a group of birds which feeds on nectar.
3 Where do butterflies have their taste sensors?
4 How does a python kill its food?
5 What is an omnivore?
6 What does an aardvark feed on?
7 Name one animal which uses a tool when feeding.
8 How does an insect-eating bat find its food?
9 Which animal feeds mainly on bamboo?
10 What do leaf-cutter ants feed on?

What's this?

This is an unusual view of a well-known animal. Can you name it?

Close up Can you name what each of these close-ups shows?

A
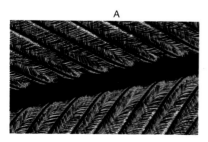

B
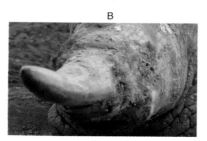

C
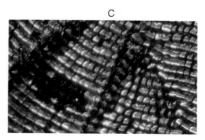

Eye spy

Which animal does each of these eyes belong to?

A

B

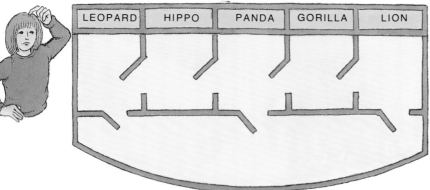

LEOPARD | HIPPO | PANDA | GORILLA | LION

Zoo puzzle

Milly got the job of keeper at her local zoo. On her first day she inspected the animals. Each cage had the animal's name printed on it, but the animals were in the wrong cages. Here is what Milly found. The lion was in the leopard's cage. The gorilla was in the hippo's cage. The hippo was in the panda's cage. The leopard was in the gorilla's cage and the panda was in the lion's cage. When Milly told the zoo owner, he told her to put each animal back in its proper cage. But he warned her that all the animals were very fierce. Before she started to move them, she was told never to put two animals in the same cage at the same time. She was also told not to put two animals in the enclosure at the same time. How many moves will Milly have to make to put each animal in its proper cage?

Mixed bag

1 What is coral made of?
2 Where would you find an aye-aye?
3 What do vultures feed on?
4 What is symbiosis?
5 What is a sundew?
6 What is the symbol of the World Wildlife Fund?
7 What is countershading?
8 Name one animal which can change color.
9 How does the African lungfish survive drought?
10 What does cold-blooded mean?
11 Where would you find a monkey-eating eagle?
12 How do flamingoes feed?
13 Which species of animal did Dian Fossey study?
14 Where do you find most marsupials?

Size

1 Name the world's largest mammal.
2 What is the world's biggest flower?
3 Name the world's smallest mammal.
4 Where does the world's biggest butterfly come from?
5 Name the world's biggest mollusk.
6 What is the world's biggest plant?
7 Name the world's biggest seed.
8 What is the world's biggest lizard?

Answers on page 152

Bookshelf

General natural history

Mysteries and Marvels of Nature series, Educational
   Development Corporation:
   Insect Life JENNIFER OWEN, 1984
   Reptile World IAN SPELLERBERG and MARIT McKERCHAR, 1984
   Bird Life IAN WALLACE, ROB HUME and RICK MORRIS, 1984
   Ocean Life RICK MORRIS, 1983
   Plant Life BARBARA CORK, 1983
   Animal world KAREN GOAMAN and HEATHER AMERY, 1983
Science Around Us series, Educational Development Corporation:
   Living Things MARIT CLARIDGE, 1985

Things to do

A Practical Guide for the Amateur Naturalist
   GERALD DURRELL with LEE DURRELL,
   Alfred A. Knopf, 1982
The Water Naturalist HEATHER ANGEL and PAT WOLSELEY,
   Facts on File, 1982
Botanic Action with David Bellamy
   CLARE SMALLMAN and DAVID BELLAMY, Hutchinson, 1978
The Curious Naturalist JOHN MITCHELL, Prentice Hall, 1980

Field Guides

The Golden Guide series, Western Publishing Company
The Peterson Field Guide series, Houghton Mifflin Company

Animals around the world

Discovering Nature series, Franklin Watts:
   *Discovering Bees and Wasps. Discovering Birds of Prey. Discovering
   Hedgehogs. Discovering Snakes and Lizards. Discovering Spiders.
   Discovering Worms*
The Birds R.T.PETERSON and THE EDITORS OF TIME-LIFE BOOKS,
   Time-Life Books, 1963

Mammals

The National Geographic Book of Mammals
The Mammals R.CARRINGTON and THE EDITORS OF TIME-LIFE
   BOOKS, Time-Life Books, 1963

Reference 1968

The Simon & Schuster Dictionary for Young Scientists
   JEANNE STONE, Simon & Schuster, 1985

Energy and life

Nature at Work: an Introduction to Ecology, British Museum (Natural
   History)/Cambridge University Press, 1978

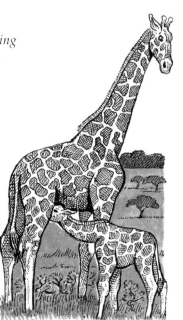

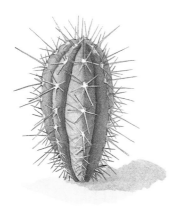

Energy and life

Nature at Work: an Introduction to Ecology, British Museum (Natural History)/Cambridge University Press, 1978

Evolution

Origin of Species, British Museum (Natural History)/Cambridge University Press, 1981

Life in cold climates

The Poles WILLY LEY and THE EDITORS OF TIME-LIFE BOOKS, Time-Life Books, 1962

Life without water

The Desert A.STARKER LEOPOLD and THE EDITORS OF TIME-LIFE BOOKS, Time-Life Books, 1961

Life between the tides

1001 Questions Answered About the Seashore N. J. BERRILL and J. BERRILL, Dover Publications, 1976

Night life

Nature's Night Life ROBERT BURTON, Blandford Books, 1982

Rain forests

Jungles Edited by EDWARD S. AYENSU, Crown Publishers, 1980

Conservation

The Doomsday Book of Animals DAVID DAY, The Viking Press, 1981

Looking for clues

Nature Detective HUGH FALKUS, Holt, Rinehart & Winston, 1979

A good read

My Family and Other Animals GERALD DURRELL, Penguin Books, 1959
The Whispering Land GERALD DURRELL, Penguin Books, 1964
A Zoo in My Luggage GERALD DURRELL, Penguin Books, 1964

Answers

page 12 The mammals.

page 18 The dachshund was bred for hunting badgers. Its name means badger hound in German. The bull-dog is short and squat with very powerful jaws because it was bred for bull baiting. This was an old English sport in which a bull was matched against several bulldogs. The dogs had to be strong and fearless.

page 19 Picasso.

page 21 The spectacled bears have different face markings because they each received different information in the sperms and eggs from their parents.

page 21 The finches have different-shaped beaks because they are adapted to eating different foods. The ground finch eats hard nuts and seeds. The warbler finch eats insects and grubs in crevices and under bark. The cactus finch probes in the flowers of cacti.

page 25 When an animal curls up into a ball it becomes a sphere. It now has a small surface compared to its volume. This means it loses less heat across its surface. This is important for hibernating mammals.

page 26 The mouse is affected more by the air than the human. It is affected less by gravity. Because of this, it is held up more by the air than the human as it falls. The mouse's relatively large surface helps it to float down more gently.

page 27 A small animal has a much bigger surface in relation to its volume than a big animal. This means it has relatively more surface to lose

heat across. Small animals, therefore, lose heat more quickly.

page 30 *toucan* fruit, insects and even eggs – mainly a herbivore but really an *omnivore*. *orangutan* fruit, leaves and some insects – mainly a herbivore but really an *omnivore*. *python* small animals such as birds and mammals – a *carnivore*. *aye-aye* fruit and insect larvae – an *omnivore* *hummingbird* nectar – a *herbivore*. *aardvark* termites – a *carnivore*. *kinkajou* plants, insects, small mammals and birds – an *omnivore*. *blue whale* small shrimps called krill – a *carnivore*.

page 51 The tongue needs to be able to feel where food is, so that it can move the food around the mouth, and to detect hard objects like fish bones. The tongue is also used to clean bits of food out of gaps between the teeth. The fingertips are also very sensitive, but they are covered with tough, protective skin which makes them less sensitive than the tongue.

page 57 No. The lantern bug lives in the forest. The caiman is much bigger and lives in the river. Predators of the lantern bug are unlikely to mistake it for a caiman.

page 60 The slope stops the honey from flowing out of the cell.

page 67 The egg has to be strong to bear the weight of the adult ostrich when the egg is being incubated. An adult ostrich weighs about 46lb, almost twice as much as a human adult.

page 68 **A** crab **B** sea urchin **C** barnacle.

page 71 Both parents take turns to incubate the eggs.

page 72 **A** happiness **B** fear **C** excitement.

page 74 Karl Hamner concluded that some animals have built-in rhythms which cannot be changed no matter what you do to the animals.

page 75 The female leatherback turtles time the return to their breeding grounds by means of their internal "clocks," which tell them when to start making their journey.

page 75 The mangrove snails reset their "clocks" every day because the times for high and low tide are different each day.

page 85 The margins of the giant water lily leaves have gaps in them to allow rainwater to drain off. Otherwise the leaf would sink under the weight of the water it collected.

page 118 Reflective mirror sides. Rest of body black. Body flattened from side to side. Upward-looking eyes. Large mouth.

page 129 Freezer store mice need thicker fur to give them better insulation. Because they are bigger than ordinary mice, they have a smaller surface area in relation to their volume compared to ordinary mice. This means they lose less heat across their surface, so they keep warmer.

page 146 *Panthera pardus* is the leopard. *Panthera onca* is the jaguar. *Panthera tigris* is the tiger.

page 146 *Panthera aquatica* or *Panthera waterus*.

page 147 **A** gentoo; **B** chinstrap; **C** rockhopper; **D** adelie; **E** king; **F** emperor.

WILDLIFE QUIZ

Heads you win: A moth **B** hippopotamus **C** giant tortoise.

True or false: 1 false; **2** false; **3** true; **4** true; **5** true; **6** false; **7** true; **8** false; **9** false; **10** false; **11** true; **12** true.

Food: 1 plankton (krill); **2** hummingbirds; **3** on their feet; **4** it constricts or strangles it; **5** an animal that eats plants and other animals; **6** termites; **7** song thrush or woodpecker finch or chimpanzee or Egyptian vulture or bearded vulture or sea otter or satin bowerbird or harvester ant or tailor ant or dwarf mongoose; **8** by sonar or echolocation; **9** giant panda; **10** fungi which they grow in special underground gardens.

What's this?: the back view of a male ostrich.

Close up: A bird's feather; **B** the horn of a rhinoceros; **C** scales of a butterfly's wing.

Eye spy: A a chameleon's eye; **B** a frog's eye.

Mixed bag: 1 limestone; **2** in Madagascar; **3** carrion (dead animals); **4** two organisms living together; **5** an insect-eating (carnivorous) plant; **6** the giant panda; **7** when an animal's body is dark on top and light underneath. The shading is sometimes reversed; **8** chameleon or plaice or octopus; **9** it "hibernates" or estivates; **10** when an animal's body temperature is always the same as its surroundings; **11** in the Philippine islands; **12** they are filter feeders; **13** the mountain gorilla; **14** in Australia.

Zoo puzzle: Milly got all the animals into the right cages in twenty-five moves. Can you do it in fewer? Ask your mathematics teacher to help.

Size: 1 the blue whale; **2** Rafflesia; **3** Savi's white-toothed pigmy shrew; **4** Papua New Guinea; **5** giant squid; **6** giant sequoia or "big tree;" **7** coco-de-mer; **8** komodo dragon.

abdomen: the hollow part of an animal's body containing most of the gut and other internal organs.

adaptation: a characteristic which improves the chances of an animal or plant's surviving and reproducing in its environment (its natural surroundings).

airfoil: a surface which is specially shaped to help it stay in the air. The wings of birds are airfoils.

air sacs: in birds, pouches formed from part of the lungs and filled with air. The air sacs make the bird very light for its size.

amphibian: an animal belonging to a group of vertebrates which live partly in water and partly on land. Frogs, toads, newts, and salamanders are amphibians. Amphibians are cold-blooded. They live mainly on land, but return to the water to breed. When young they breathe with gills, but as adults they use their lungs. Adult amphibians usually have 4 legs and smooth, wet skins.

antennae: the pairs of feelers found on the heads of insects and other arthropods. Antennae are used for touch, taste, smell and for sensing changes in humidity and temperature.

aquatic: describes animals or plants that live in water.

binocular vision: the use of both eyes to look at the same object. Binocular vision is possible if the eyes are at the front of the head. Because each eye looks at the object from a different angle, two different pictures are seen. From this the brain can work out how far away the object is. Binocular vision is important for animals which need to judge distances for hunting or for movement.

bladder: a stretchy, baglike structure.

breeding ground: the place to which animals travel to breed and bring up their young.

breeding program: the carefully planned arrangements made to encourage animals in captivity to breed.

bulb: a short underground stem surrounded by fleshy leaves swollen with stored food.

buoyancy: the tendency for some objects to float or rise when submerged in a fluid.

camouflage: a form of disguise which helps an animal to blend with its background so that it is not noticed by other animals.

carbon dioxide: a gas made up of the elements carbon and oxygen. Carbon dioxide is present in the air, and is also dissolved in the water of rivers, lakes and oceans.

carcass: the dead body of an animal.

carnivore: a flesh-eating animal.

cell: the basic unit of living matter. It contains a jellylike living material, called protoplasm, surrounded by a thin skin called the membrane. Plant cells also have a stiff cell wall around the outside of each cell.

cellulose: the material that forms the walls of plant cells.

characteristic: any shape, pattern, or way of behaving by which an organism can be recognized.

chitin: a hard, horny, waterproof substance found in the shells of crustaceans and the hard outer covering of insects and other arthropods.

chlorophyll: the green substance found in plants. It absorbs light, which the plant uses to make its own food by photosynthesis.

chrysalis: *see* **pupa.**

class: one of the groups used in the classification of plants and animals. Mammalia is the class to which the mammals belong.

classification: the arrangement of animals and plants in groups, chosen by looking at their characteristics.

cocoon: a fluffy ball of silk made by an animal. Some moth caterpillars spin cocoons around themselves when they are ready to turn into moths.

cold-blooded: describes an animal whose body temperature changes as the temperature of its surroundings changes. In cold weather, the animal's temperature falls, but in hot weather the body temperature rises, so its blood is not always cold.

colony: a group of organisms of the same species living together.

compound eye: an eye made up of many tiny units, each with its own lens.

conifers: trees and shrubs which produce cones.

copepod: microscopic crustaceans found in very large numbers in both fresh and salt water.

corm: a swollen underground stem which stores food and produces new plants from its bulbs.

countershading: a form of camouflage where the animal is colored dark on top and lighter underneath.

courtship: a complex pattern of behavior and signaling which takes place before mating.

crustacean: one of a group of hard-skinned animals with jointed legs and long antennae. The body is divided into three main parts, which may be further divided into segments. Crustaceans include crabs, shrimps, lobsters, barnacles, water fleas and copepods.

cuticle: a waterproof layer covering the outside of a living organism.

decomposer: a living organism which causes decay, breaking down the remains of dead plants or animals.

diurnal: describes an animal which is active only during the day.

dormant: describes an organism which is resting and not growing.

drag: the resistance an animal meets as it tries to move through the air or water.

drift line: the line marking the highest point the sea reaches on a beach at high tide. It is marked by an untidy line of drying seaweed, empty shells, driftwood and other objects washed up by the tide.

echolocation: a method of navigation using echoes. High-pitched sounds are sent out, and these sounds bounce back (echo) off solid objects in their path.

egg: a female sex cell both before and after it has been fertilized. Eggs are also the hard- or soft-shelled structures in which the embryos of some animals develop.

endangered species: species which have so few members that they are in danger of becoming extinct.

environment: the collection of physical, chemical and biological factors which affect an organism.

estivation: a kind of hibernation or dormancy shown by some desert creatures during the dry season.

evolution: the way organisms change over many generations, giving rise to new species.

extinct: describes a species that has died out and no longer exists.

fertilization: the joining together of a male sex cell and a female sex cell (in animals, these are called sperm and egg) to produce a new living organism.

filter feeder: an animal which feeds by sieving out tiny particles of dead or living organic material from water as it flows through or past the animal.

fluke: the horizontal tail fin of a whale or dolphin.

food chain: a chain which shows the order in which food energy is passed from plants to animals. All food chains begin with a plant. This is eaten by an animal which is eaten by another animal and so on.

food web: a number of food chains linked together to form a web.

genus: one of the groups used in the classification of plants and animals. *Pinus* is the genus to which pine trees belong.

germination: the growth and development of a plant embryo, in a seed or spore, into an independent plant.

gestation time: the length of time that a female mammal carries her young inside her body. It starts when the egg is fertilized and ends when the young are born.

gills: structures used for breathing under water.

habitat: the place where a particular organism lives.

herb: a nonwoody flowering plant.

herbivore: an animal that feeds mainly on plants.

hibernate: to go into a very deep sleep during cold weather.

hormones: chemicals produced by living organisms. They are used to coordinate body processes, such as growth and reproduction.

humidity: a measure of the amount of moisture in the atmosphere.

incubate: to keep eggs warm so that they will hatch.

insectivore: an animal that feeds mainly on insects.

insulate: to prevent heat from leaving or entering an object or animal.

internal "clock": many living organisms have a built-in way of recognizing the time of day or the time of year. We say that they have an internal clock.

invertebrate: an animal without a backbone.

larva (pl. larvae): a young animal that often looks quite different from its parents. Tadpoles are the larvae of frogs.

lichen: a small, plantlike growth found on trees and bare rocks. It is a mixture of an alga and a fungus.

lift: the lifting force which develops because of a difference in air pressure above and below the wings of a bird or an airplane.

locomotion: movement from place to place. Walking, flying and swimming are types of locomotion.

magnetic field: the area around a magnet where its magnetic force acts. The Earth is rather like a giant magnet.

mammal: a backboned animal which is warm-blooded and whose body is usually covered with hair or fur. The female mammal gives birth to live young and feeds them on milk.

marsupial: a pouched animal.

mass: the amount of material in a body or object.

metamorphosis: the change of shape in some animals during their life cycle.

microscopic: describes something so small that a microscope is needed to see it.

migration: the movement of animals from one place to another, often over long distances.

mimicry: copying the appearance, movement or sound of another species.

mollusk: a member of a large group of animals, without a backbone and with a soft body with a muscular foot. Many mollusks are protected by a hard shell.

mucus: a slimy fluid produced by animals. Mucus moistens and protects delicate surfaces and prevents their drying out.

natural selection: in evolution, the gradual process during which some species survive and reproduce while others die out and become extinct.

navigation: the method used by an animal to find its way about.

nectar: a sweet liquid made by some flowers.

nocturnal: describes an animal or plant that is active at night.

nymph: one of the stages in the development of some insects before they become adults. Nymphs look similar to their parents but are smaller.

omnivore: an animal which eats plants and other animals.

order: one of the groups in the classification of animals and plants. It is a unit into which classes are divided. An order is divided up further into families.

organ: a distinct part of an animal or plant made of many specialized cells. Each organ has its own special job to do.

organic: describes any material which contains carbon. Carbon compounds are found in all living and dead cells.

oxygen: a gas that is found in air and water. It has no color, taste or smell. It is necessary for plant and animal life.

parasite: an animal or plant which lives on, or inside, another organism (called the host). It gets food from its host, but the host suffers as a result.

pheromone: a chemical substance produced by animals which acts as a signal.

photosynthesis: a process by which green plants use energy from sunlight to make sugars from carbon dioxide and water.

plankton: microscopic animals and plants which float near the surface of the sea and freshwater lakes.

plumage: the feathers of a bird.

pollen: tiny yellow or orange grains produced by the male parts of flowering and cone-bearing plants. The pollen grains contain the male sex cells.

pollination: the transfer of pollen grains from the male parts to the female parts of a flowering plant.

population: the total number of organisms of a species living in a particular area at any one time.

polyp: a hollow, cup-shaped animal with a ring of tentacles round its mouth.

predator: an animal that hunts and eats other animals.

prey: an animal that is hunted and eaten as food by another animal.

primate: a group of mammals which includes monkeys, apes and humans.

prop root: a special type of root produced by some plants which helps support the main stem or trunk.

proteins: a group of complicated chemical compounds containing nitrogen. Proteins are important for plant and animal growth.

pupa: a stage in the life cycle of an insect during which the animal changes from a caterpillar to an adult insect. Usually the pupa is covered in a hard case. Inside, the body materials of the caterpillar are rearranged to form the adult butterfly or moth. The pupa of a moth or butterfly is sometimes called a chrysalis.

reproduction: the process by which living organisms produce offspring.

reptile: one of a large group of cold-blooded animals with backbones and a dry, scaly skin. Most reptiles live on land and lay eggs covered with a tough, leathery shell.

rhythm: an action which occurs at regular intervals.

scavenger: an animal which feeds on dead animal or plant material.

sensor: a group of cells which picks up signals from its surroundings.

sex cell: a special kind of cell produced when animals or plants reproduce. Sex cells are either male or female. In animals, male sex cells are called sperm. Female sex cells are called eggs. Male and female sex cells join together during fertilization.

single lens reflex camera: a type of camera which allows you to view the object you want to photograph through the camera's lens.

siphon: a tubelike structure that draws water into an animal.

sonar: a method of navigation using round. High-pitched sounds are given out and they bounce back off any objects in their path.

species: a unit of classification of animals and plants. Members of one species can breed among themselves. They cannot usually breed with members of another species.

sperm: a male reproductive/sex cell.

spore: a special reproductive unit produced in large numbers by certain plants and other organisms. Spores are carried by wind and air currents.

stoma (pl. stomata): one of the many tiny pores or holes found in the outer layer of leaves and also on some plant stems.

streamlined: describes a shape that is pointed or rounded at the front end and tapers to a thinner shape at the back.

submerged: describes an object that is under the surface of a liquid such as water.

suckle: to feed a young mammal with milk from the special milk-producing glands. All female

mammals suckle their young for the first part of their lives.

surface film *see* **surface tension.**

surface tension: the property of liquids that makes their surface appear to be covered by a thin, elastic film, called the surface film.

symbiosis: the living together of two different types of organism.

territory: an area of land or water in which an animal or a group of animals live, feed and breed.

thermostat: a device for controlling the temperature of an object.

torpor: the condition of being very sleepy and inactive.

tuber: an underground stem swollen with food.

ultraviolet light: invisible rays present in sunlight.

umbilical cord: a soft cord of living tissue that joins the developing baby mammal to its mother before birth.

underexposing: in photography, not allowing enough light when taking a picture.

variation: differences between animals or plants of the same species.

vertebrate: any animal which has a skeleton of bone or cartilage. Fish, amphibians, reptiles, birds and mammals are all vertebrates.

warm-blooded: describes an animal which has a body temperature that stays the same whatever the temperature of the surroundings. Birds and mammals are warm-blooded.

weaned: describes a young mammal which has stopped feeding on its mother's milk and has started eating other food.

womb: the part of a female mammal's body where the baby develops before birth.

zooplankton: microscopic animals which float at the surface of the sea and freshwater lakes.

Index

INDEX